SpringerWienNewYork

CISM COURSES AND LECTURES

The series presents lecture notes, monographs, edited works and proceedings in the field of Mechanics, Engineering, Computer Science and Applied Mathematics.
Purpose of the series is to make known in the international scientific and technical community results obtained in some of the activities organized by CISM, the International Centre for Mechanical Sciences.

INTERNATIONAL CENTRE FOR MECHANICAL SCIENCES

COURSES AND LECTURES - No. 509

MIXED FINITE ELEMENT TECHNOLOGIES

EDITED BY

CARSTEN CARSTENSEN
HU BERLIN, GERMANY

PETER WRIGGERS
UNIVERSITY OF HANNOVER

SpringerWienNewYork

This volume contains 55 illustrations

ISBN 978-3-211-99950-9
e-ISBN 978-3-211-99094-0

PREFACE

The finite element method is the dominating tool in computational mechanics. To quote the late Simo: "Mechanics without finite elements is like rock'n roll without electricity". In straightforward finite element schemes, there is an elastic energy which is minimized over some conforming discrete subspace X_h. *This leads to a linear system of equations of the form*

$$a(u_h, v_h) = b(v_h) \qquad \text{for all } v_h \in Y_h$$

with unknown discrete solution u_h *in* X_h *for a given right-hand side* $b(v_h)$ *and the bilinear form* $a(u_h, v_h)$. *In the context of an elastic energy* $\frac{1}{2}a(v_h, v_h) - b(v_h)$, *the bilinear form is symmetric and one choice for trial and test functions reads* $X_h = Y_h$.

In saddle-point problems, which are related to different engineering applications in mechanics, a mixed finite element method may consider the general situation where $X_h \neq Y_h$. *This is the starting point of the more involved mathematical justification of the mixed finite element method. The choice of trial and test functions* X_h *and* Y_h *is subject to an inf-sup condition named after Babuska, Brezzi, Ladyzhenskaya in order to guarantee stability.*

Therefore, the use of arbitrary discrete spaces - often employed in engineering finite element analysis - fails in general. More often even in everydays practise, there are methods applied which "work on Mondays but not on Tuesdays" (saying after R. Stenberg). Hence it is mandatory to consult some mathematics in order to construct reliable and fast mixed finite element simulation tools. This CISM course brought together leading experts in the field of nonstandard finite element methods to highlight the state of the art on mathematical and engineering aspects of current mixed finite element technology. The basic lectures on the necessary mathematical and mechanical background for linear and nonlinear application areas of mixed finite elements methods were given by the two editors and followed by particular applications and methods discussed by the other lecturers.

Carsten Carstensen and Peter Wriggers

CONTENTS

Lectures on Adaptive Mixed Finite Element Methods*

C. Carstensen[†]

[†]Department of Mathematics, Humboldt-Universität Berlin, Germany

Abstract These lectures concern the three most simple model problems of elliptic second order partial differential equations which allow for some mixed formulations. The introduction to the Poisson problem, to the Stokes problem, and to linear elasticity is completely standard and hence kept rather short.

The first aim is a general discussion of the mixed formulations around various statements of the inf-sup condition often named after Ladyzhenskaya, Babuška, Brezzi. Some details on the implementation of Raviart-Thomas mixed finite elements in MATLAB complement this introduction.

The second aim is a particular outline of the author's own research on a posteriori error analysis and adaptive algorithms of mixed finite element methods displayed for the Poisson problem. The presentation is motivated by the author's research Braess et al. (2004); Carstensen (1997, 1999, 2005); Carstensen and Dolzmann (1998); Carstensen et al. (2000); Carstensen and Funken (2001a,b); Carstensen and Verfürth (1999) with essential help of many researchers including S. Bartels, D. Braess, G. Dolzmann, S.A. Funken and R. Hoppe.

These lecture notes are written with the help of Wolfgang Boiger, Andreas Byfut, David Günther, Athina Konstantinidou, Hella Rabus, and Jan Reininghaus. The support of all these scientists, especially that of the younger ones is thankfully acknowledged.

1 Three Mixed Formulations

This lecture introduces mixed formulations for three model examples and displays them in a unified abstract way, which the remaining lectures are based on.

*Supported by the DFG Research Center MATHEON "Mathematics for key technologies" in Berlin.

1.1 Mixed Formulations

If the main interest is on an accurate stress or flux approximation and some strict equilibration condition, it might be advantageous to consider an operator split: Instead of one partial differential equation of order $2m$ one considers two equations of order m.

To be more precise, given one equation in an abstract form $\mathcal{L}u = G$ with some differential operator $\mathcal{L} = \mathcal{AB}$ composed of $\mathcal{A}$ and $\mathcal{B}$, define $p := \mathcal{B}u$ and solve the two equations

$$\mathcal{A}p = G \quad \text{and} \quad \mathcal{B}u = p. \tag{1}$$

Throughout these lecture notes, a and b are two bilinear forms associated with $\mathcal{A}$ and $\mathcal{B}$ on the Hilbert spaces L and H with dual spaces L^* and H^*, typically, L is some Lebesgue space and H is some Sobolev space. Given bounded bilinear forms $a : H \times H \to \mathbb{R}$, $b : H \times L \to \mathbb{R}$ and right-hand sides $g \in L^*$, $f \in H^*$, a weak form of (1) reads

$$\begin{aligned} a(p,q) + b(q,u) &= f(q) \quad \text{for all} \quad q \in H; \\ b(p,v) &= g(v) \quad \text{for all} \quad v \in L. \end{aligned} \tag{2}$$

Note that $p - \mathcal{B}u = f$ represents the strong form of $(2)_a$ and $\mathcal{A}p = G$ is the strong form of $(2)_b$.

1.2 Lebesgue and Sobolev Spaces

The reader is expected to be familiar with the basic features of Lebesgue and Sobolev functions. For the understanding of these lecture notes, it suffices to know that $v \in L^2(\Omega)$ for some domain $\Omega \subset \mathbb{R}^n$ means that v is measurable and the (Lebesgue) integral of $|v(x)|^2$ over Ω is finite,

$$\|v\|^2_{L^2(\Omega)} := \int_\Omega |v(x)|^2 \, dx < \infty.$$

Then, this defines the norm of v in $L^2(\Omega)$. The same notation applies to more components such as $L^2(\Omega; \mathbb{R}^m)$ and matrices $L^2(\Omega; \mathbb{R}^{m \times n})$, where each component belongs to $L^2(\Omega)$.

For a smooth function v, the Sobolev norm $\| \cdot \|_{H^1(\Omega)}$ is defined by

$$\|v\|^2_{H^1(\Omega)} := \int_\Omega |v(x)|^2 \, dx + \int_\Omega |\nabla v(x)|^2 \, dx$$

and $H^1(\Omega)$ denotes the completion of smooth functions with respect to this norm. To be more precise, let $C_c^\infty(\Omega)$ denote the set of arbitrarily smooth functions with a compact support inside Ω. Then, the Sobolev space $H_0^1(\Omega)$ is defined as the closure of $C_c^\infty(\Omega)$ with respect to the Sobolev norm $\|\cdot\|_{H^1(\Omega)}$, while $H^1(\Omega)$ is the closure of $C_c^\infty(\mathbb{R}^n)$ restricted to Ω under the same norm. One says that $v \in H^1(\Omega)$ satisfies $v|_{\partial\Omega} = 0$ in the sense of traces, written $\gamma(v) = 0$, if and only if $v \in H_0^1(\Omega)$.

It can be proven that Sobolev functions (i.e., the elements in such a Sobolev space) have a weak derivative in $L^2(\Omega; \mathbb{R}^n)$ and satisfy the integration by parts formula: For all $u, v \in H^1(\Omega)$ and $j = 1, \dots, n$ there holds

$$\int_\Omega u \frac{\partial v}{\partial x_j}\, dx + \int_\Omega \frac{\partial u}{\partial x_j}\, v\, dx = \int_{\partial\Omega} \gamma(u)\, \gamma(v)\, \nu_j\, ds. \tag{3}$$

Here ν is the outer unit normal vector along the boundary $\partial\Omega$ and the trace operator $\gamma : H^1(\Omega) \to L^2(\partial\Omega)$ is well defined, linear, continuous, and satisfies $(\gamma u)(x) = u(x)$ for all $x \in \partial\Omega$ and all $u \in C(\overline{\Omega})$. An alternative notation reads $\gamma(u) = u|_{\partial\Omega}$ or one simply replaces $\gamma(u)$ by u on the right-hand side of (3).

Moreover, $H(\mathrm{div}, \Omega) \subset L^2(\Omega)^n$ is the set of functions whose divergence is integrable. The norm in $H(\mathrm{div}, \Omega)$ is given by

$$\|v\|^2_{H(\mathrm{div},\Omega)} = \int_\Omega |v(x)|^2\, dx + \int_\Omega |\operatorname{div} v(x)|^2\, dx.$$

Note that $H^1(\Omega)^n \subset H(\mathrm{div}, \Omega) \subset L^2(\Omega)^n$, but $H^1(\Omega)^n \neq H(\mathrm{div}, \Omega) \neq L^2(\Omega)^n$ for $n > 1$. Finally,

$$H^k_{\mathrm{loc}}(\Omega) = \{V : \Omega \to \mathbb{R} \text{ measurable} : \forall K \subset\subset \Omega,\ v|_K \in H^k(\Omega)\},$$

where $K \subset\subset \Omega$ denotes an open bounded subset with $\overline{K} \subset \Omega$.

1.3 Poisson Problem

The stationary heat equation and many mathematical models in applications of solid and fluid mechanics lead to the Laplace operator $\Delta = \frac{\partial^2 \cdot}{\partial x_1^2} + \dots + \frac{\partial^2 \cdot}{\partial x_1^n}$. Given $g \in L := L^2(\Omega)$, let $u \in H^1(\Omega)$ with $p := \nabla u \in H := H(\mathrm{div}, \Omega)$ denote the solution to the *Poisson problem*

$$\Delta u + g = 0 \text{ in } \Omega \quad \text{and} \quad u = 0 \text{ on } \partial\Omega. \tag{4}$$

The weak form of (4) is given by (2) with $p, q \in H$, $u, v \in L$, and (where $\cdot$ denotes the scalar product in $\mathbb{R}^n$)

$$a(p,q) := \int_\Omega p \cdot q\, dx, \quad b(q,v) := \int_\Omega v \operatorname{div} q\, dx,$$

$$f(q) := 0, \qquad g(v) := -\int_\Omega g\, v\, dx.$$

It is known that there exists a unique solution and, at least locally, $u \in H^2_{\mathrm{loc}}(\Omega)$ and $p \in H^1_{\mathrm{loc}}(\Omega)$.

1.4 Stokes Problem

The stationary incompressible fluid flow can be modeled by the Stokes equations for given $f \in L^2(\Omega)$ and unknown velocity u and pressure p,

$$\Delta u + \nabla p = -f \text{ in } \Omega, \quad \operatorname{div} u = 0 \text{ in } \Omega, \quad u = 0 \text{ on } \partial\Omega. \tag{5}$$

The weak form of (5) is given by (2) with $p, q \in L := L^2_0(\Omega)$ and $u, v \in H := H^1_0(\Omega)^n$ and, (where : denotes the scalar product in $\mathbb{R}^{n\times n}$)

$$a(u,v) := \int_\Omega \nabla u : \nabla v \, dx, \quad b(v,q) := \int_\Omega q \operatorname{div} v \, dx,$$
$$f(v) := \int_\Omega f \, v \, dx, \qquad g(q) := 0.$$

Here, in comparison to (2), the roles of u and p are interchanged, which typically leads to some confusion. Moreover, $L^2_0(\Omega) := \{q \in L^2(\Omega) : \int_\Omega q \, dx = 0\}$ fixes a global additive constant in the pressure (because of lacking Neumann boundary conditions). Then there exists a unique solution (u, p).

1.5 Elasticity Problem

We adopt the notation of the previous two subsections and continue with a linear stress-strain relation with the linear Green strain $\varepsilon(u) := \operatorname{sym}(\nabla u) := (\nabla u + (\nabla u)^T)/2$ of the form

$$g + \operatorname{div} \mathbb{C}\,\varepsilon(u) = 0 \quad \text{in } \Omega \quad \text{and} \quad u = 0 \text{ on } \partial\Omega. \tag{6}$$

The weak form of (6) is given by (2) with $\sigma, \tau \in H := H(\operatorname{div}, \Omega)^n \cap L^2(\Omega; \mathbb{R}^{n\times n}_{\mathrm{sym}})$, $u, v \in L := L^2(\Omega)^n$, and

$$\begin{aligned} a(\sigma,\tau) &:= \int_\Omega (\mathbb{C}^{-1}\sigma) : \tau \, dx, \quad & b(\tau,v) &:= \int_\Omega \operatorname{div} \tau \cdot v \, dx, \\ f(\tau) &:= 0, & g(v) &:= -\int_\Omega g \, v \, dx. \end{aligned} \tag{7}$$

Here, $\mathbb{C}$ denotes the linear fourth-order tensor, namely

$$\mathbb{C}A := \lambda \operatorname{tr}(A)\, \mathbf{1} + 2\mu\, A \quad \text{for } A \in \mathbb{R}^{n\times n},$$

with inverse relation

$$\mathbb{C}^{-1}A = 1/(2\mu)\, A - \lambda/(2\mu(n\lambda + 2\mu)) \operatorname{tr}(A)\, \mathbf{1} \quad \text{for } A \in \mathbb{R}^{n\times n}.$$

The material parameters λ and μ are positive and hence (6) is an elliptic PDE with a unique solution $u \in H_0^1(\Omega) \cap H_{\mathrm{loc}}^2(\Omega)$.

2 First Analysis for Mixed Formulations

Let X, Y be real Hilbert spaces with duals X^*, Y^* and let $a : X \times Y \to \mathbb{R}$ be a bounded bilinear form, i.e., for all $u, w \in X, v, z \in Y$ there holds

$$a(\lambda\, u, \mu\, v) = \lambda\, \mu\, a(u,v) \text{ for all } \lambda, \mu \in \mathbb{R}, \qquad \text{(homogenity)}$$
$$a(u+w, v+z) = a(u,v) + a(u,z) + a(w,v) + a(w,z), \qquad \text{(additivity)}$$
$$\|a\| := \sup_{\substack{x\in X\\ \|x\|=1}} \sup_{\substack{y\in Y\\ \|y\|=1}} a(x,y) < \infty. \qquad \text{(boundedness)}$$

Given such a bounded bilinear form $a : X \times Y \to \mathbb{R}$, one defines the two associated linear operators

$$\begin{aligned} &A_1 \in L(X;Y^*) \quad \text{and} \quad A_2 \in L(Y;X^*) \quad \text{by} \\ &A_1 x := a(x,\cdot) \quad \text{and} \quad A_2 y := a(\cdot,y) \quad \text{for } x \in X \text{ and } y \in Y. \end{aligned} \tag{8}$$

The operator norms of A_1 and A_2 read

$$\|A_1\| := \|A_1\|_{L(X,Y^*)} = \sup_{x\in X\setminus\{0\}} \sup_{y\in Y\setminus\{0\}} \frac{a(x,y)}{\|x\|_X \|y\|_Y} = \|A_2\|_{L(Y,X^*)} = \|a\|.$$

Remark 2.1. Conversely, any $A_1 \in L(X;Y^*)$ defines some bilinear form $a : X \times Y \to \mathbb{R}$ which is bounded with $\|a\| = \|A_1\|$. Analogous remarks apply to A_2.

An operator A_1 (resp. A_2) from (8) is called an isomorphism if it has a bounded inverse so that the linear equation $A_1 x = f$ (resp. $A_2 y = g$) has a unique solution, which is bounded with respect to the data f (resp. g), in other words, the problem is *well posed.* Sufficient and necessary conditions for an isomorphism A_j are well known Babuška and Azis, Brezzi and Fortin (1991). The proof is sketched for convenient reading to narrow the gap between lectures in functional analysis and on mixed FE formulations.

Theorem 2.2 (Invertibility of $A_j \Leftrightarrow$ Inf-Sup Condition)**.** *Let X, Y be Hilbert spaces and let $a : X \times Y \to \mathbb{R}$ be a bounded bilinear form with A_1 and A_2 as in* (8).

a. The following conditions (i),(ii),(iii),(iv) are pairwise equivalent

i) $\forall f \in Y^* \; \exists!\, x \in X, \quad a(x,\cdot) = f;$

ii) $\forall g \in X^* \; \exists!\, y \in Y, \quad a(\cdot,y) = g;$

iii) $\alpha_1 := \inf_{\substack{x\in X\\ \|x\|_X=1}} \sup_{\substack{y\in Y\\ \|y\|_Y=1}} a(x,y) > 0$ *and*

$\forall y \in Y \setminus \{0\} \; \exists x \in X, \quad a(x,y) \neq 0;$

iv) $\alpha_2 := \inf_{\substack{y \in Y \\ \|y\|_Y = 1}} \sup_{\substack{x \in X \\ \|x\|_X = 1}} a(x,y) > 0$ *and*

$\forall x \in X \setminus \{0\} \; \exists y \in Y, \quad a(x,y) \neq 0.$

b. There holds $A_1^* = A_2$, $A_2^* = A_1$ *and each of the above conditions (i),(ii),(iii),(iv) implies that* A_1 *and* A_2 *are continuously invertible and*

$$\|A_1^{-1}\|_{L(Y^*;X)} = \frac{1}{\alpha_1} = \frac{1}{\alpha_2} = \|A_2^{-1}\|_{L(X^*;Y)} \,.$$

Proof of Theorem 2.2.
Claim 1: If $\alpha_1 > 0$, then Range(A_1) is a closed subspace of Y^*.

Proof. Suppose (f_j) is a Cauchy sequence in Range(A_1). There exists a sequence (x_j) in X with $f_j = A_1 x_j$ for all $j \in \mathbb{N}$. By definition of $\alpha_1 > 0$ there holds, for all $j, k \in \mathbb{N}$,

$$\alpha_1 \|x_j - x_k\|_X \le \|A_1 x_j - A_1 x_k\|_{Y^*} = \|f_j - f_k\|_{Y^*}.$$

Hence (x_j) is also a Cauchy sequence in the complete space X, and so is convergent towards some x. Since A_1 is bounded,

$$(f_j) = (A_1 x_j) \longrightarrow f := A_1 x \quad \text{in } Y^*.$$

Hence $(f_j) \to f$ has a limit in Range(A_1). Therefore, Range(A_1) is complete, whence closed.

Claim 2: If Range(A_1) closed and $\ker(A_2) = \{0\}$, then Range$(A_1) = Y^*$.

Proof. Let $R_2 : Y^* \to Y$ be the Riesz representation in the Hilbert space Y, i.e., for each $f \in Y^*$, $z := R_2 f \in Y$ satisfies

$$\langle z, y \rangle_Y = f(y) \quad \text{for all } y \in Y.$$

Suppose Range$(A_1) \subsetneq Y^*$. Then $(R_2 \,\text{Range}(A_1)) \subsetneq R_2 Y^* = Y$ and there exists some $y \in Y \setminus \{0\}$ such that $y \perp (R_2 \,\text{Range}(A_1))$. Hence, for all $x \in X$,

$$0 = \langle R_2 A_1 x, y \rangle_Y = (A_1 x)(y) = a(x,y) = (A_2 y)(x).$$

This reads $y \in \ker(A_2) = \{0\}$ and contradicts $y \neq 0$.

Proof of (iii) $\Rightarrow$ (i): This follows from Claim 1 and 2.

Proof of (i) ⇒ (iii): The inverse mapping theorem applies to $A_1 : X \to Y^*$ and shows

$$0 < ||A_1^{-1}||_{L(Y^*;X)}^{-1} = \left(\sup_{\substack{f\in Y^* \\ ||f||_{Y^*}=1}} ||A_1^{-1}f||_X \right)^{-1} = \inf_{\substack{f\in Y^* \\ ||f||_{Y^*}=1}} ||A_1^{-1}f||_X^{-1}$$

$$= \inf_{f\in Y^*\setminus\{0\}} \frac{||f||_{Y^*}}{||A_1^{-1}f||_X} = \inf_{x\in X\setminus\{0\}} \frac{||A_1\,x||_{Y^*}}{||x||_X} = \alpha_1.$$

This proves the first assertion in (iii) and $1/\alpha_1 = ||A_1^{-1}||_{L(Y^*;X)}$ in (b). The second assertion in (iii) is verified as follows: Given any $y \in Y \setminus \{0\}$, a corollary to the Hahn-Banach theorem guarantees the existence of some $f \in Y^*$ with $f(y) \neq 0$. From (i), there exists $x \in X \setminus \{0\}$ with $f = A_1\,x$. Hence, $a(x,y) = (A_1\,x)(y) = f(y) \neq 0$. This completes the proof of (iii).

Since $A_1^* \in L(Y;X^*)$ satisfies, for all $x \in X$ and $y \in Y$, that

$$(A_1^*\,y)(x) = A_1(x)(y) = a(x,y) = (A_2\,y)(x),$$

there holds $A_1^* = A_2$. If A_1 is invertible, so is its dual A_2 and vice versa. This proves $(i) \Leftrightarrow (ii)$. Moreover, the norms of A_2^{-1} and A_1^{-1} coincide. Hence $\alpha_1 = \alpha_2$. This indicates the proof of (b). The remaining details and implications in the proof of (a) are omitted. □

Definition 2.3. Some bilinear form $a : X \times X \to \mathbb{R}$ is called X−elliptic if there exists $\alpha > 0$ such that, for all $x \in X$, there holds

$$\alpha\,||x||_X^2 \leq a(x,x).$$

Example 2.4. Suppose $a : X \times X \to \mathbb{R}$ is bounded and X-elliptic. Clearly, $\alpha \leq \alpha_1$ in (iii) and, by Theorem 2.2, A_1 is an isomorphism. This statement is known as the generalised Lax-Milgram lemma and plays a dominant role in the existence theory of elliptic PDEs.

Theorem 2.5 (Main Theorem for Mixed Formulations). *Let X,Y be Hilbert spaces and let $a : X \times X \to \mathbb{R}$ and $b : X \times Y \to \mathbb{R}$ be bounded bilinear forms. Set*

$$V := \ker(B_1) = \{x \in X \,:\, b(x,\cdot) = 0 \text{ in } Y^*\}$$

and suppose that

$$0 \leq \alpha := \inf_{\substack{x\in V \\ ||x||_X=1}} \sup_{\substack{u\in V \\ ||u||_X=1}} a(x,u) \leq ||a|| < \infty,$$

$$0 \leq \beta := \inf_{\substack{y\in Y \\ ||y||_Y=1}} \sup_{\substack{x\in X \\ ||x||_X=1}} b(x,y) \leq ||b|| < \infty.$$

Then the linear continuous map

$$L: \begin{cases} X \times Y \to (X \times Y)^*, \\ (x,y) \mapsto a(x,\cdot) + b(\cdot,y) + b(x,\cdot) \end{cases}$$

is an isomorphism if and only if

$$\alpha, \beta > 0 \text{ and } \forall\, y \in V \setminus \{0\}\ \exists\, x \in V, \quad a(x,y) \neq 0\,.$$

Remark 2.6. The summands in $a(x,\cdot) + b(\cdot,y) + b(x,\cdot)$ act on different spaces according to their domain, such that, in more details,

$$L(x,y)(u,w) := a(x,u) + b(u,y) + b(x,w).$$

Remark 2.7. Since X, Y are Hilbert spaces, infima und suprema are attained in the definition of α, β, etc. For example, $\operatorname{dist}(x,V) = \|x - v\|$ for some $v \in V$. In fact, for a minimal sequence (x_j) in V we have that $\|x_j\| \leq \|x - x_j\| + \|x\|$ is bounded and there is a subsequence $(x_k) \rightharpoonup v$ in V. Because $\|x - \cdot\|$ is convex and continuous, it is weakly lower semi continuous and so $\|x - v\| \leq \lim_{k\to\infty} \|x - x_j\| = \operatorname{dist}(x,V)$.

Remark 2.8. The condition $\beta > 0$ for some bilinear form $b(\cdot,\cdot)$ is called LBB-condition (written 'b satisfies (LBB)') after Ladyzhenskaya, Babuška, and Brezzi. However, the notion of an LBB condition is more often applied to the discrete situation, cf. Definition 3.1.

Proof of Theorem 2.5.

"$\Longrightarrow$" Consider $L: X \times Y \to (X \times Y)^*$ with $(x,y) \mapsto a(x,\cdot) + b(\cdot,y) + b(x,\cdot)$ which satisfies the inf-sup condition with a constant $\gamma > 0$. Then, for all $y \in Y$ there exists $(u,w) \in X \times Y$ such that

$$\gamma \, \|y\|_Y \, \|u\|_X \leq \gamma \, \|(0,y)\|_{X\times Y} \, \|(u,w)\|_{X\times Y} \leq b(u,y).$$

This implies $\beta \geq \gamma > 0$. Hence, the restriction of B_1 to $V^\perp := \{x \in X : x \perp V\}$, namely

$$B: V^\perp \to Y^*, \quad x \mapsto b(x,\cdot),$$

is an isomorphism by Theorem 2.2, since (iv) and hence (a)-(b) hold with

$$\inf_{\substack{x \in V^\perp \\ \|x\|_X = 1}} \ \sup_{\substack{y \in Y \\ \|y\|_Y = 1}} b(x,y) = \beta > 0.$$

It remains to show that

$$A: V \to V^*, \quad v \mapsto a(v, \cdot)|_V$$

is an isomorphism, as well. Once this is established, the remaining part of the claim follows from Theorem 2.2 applied to $a|_{V\times V}$.

Let $f \in V^*$. By the Hahn-Banach theorem, there exists a continuation $F \in X^*$ of f. Since L is isomorphic, there exists $(x, y) \in X \times Y$ such that

$$L(x, y)(u, w) = a(x, u) + b(u, y) + b(x, w) = F(u)$$

for all $(u, w) \in X \times Y$. This expression does not depend on w and so $b(x, \cdot) = 0$, whence $x \in V$. For $u \in V$, this leads to

$$a(x, u) = F(u) = f(u),$$

and so to $Ax = a(x, \cdot)|_V = f$. Consequently, A is surjective.

Let $x \in \ker(A) \subset V$, i.e., $a(x, v) = 0$ for all $v \in V$. Since B is an isomorphism the adjoint operator

$$B^*: Y \to (V^\perp)^*, \quad y \mapsto b(\cdot, y)|_{V^\perp}$$

is an isomorphism, as well. Hence there exists a unique $y \in Y$ with

$$b(\cdot, y)|_{V^\perp} = -a(x, \cdot)|_{V^\perp} \in (V^\perp)^*.$$

Since $b(\cdot, y)|_V = 0 = -a(x, \cdot)|_V$, there holds

$$b(\cdot, y) = -a(x, \cdot) \in X^*.$$

This and $x \in V$, i.e., $b(x, \cdot) = 0$, implies

$$L(x, y) = a(x, \cdot) + b(\cdot, y) + b(x, \cdot) = 0.$$

By assumption, L is isomorphic and so $(x, y) = (0, 0)$. Hence, $x \in \ker(A)$ implies $x = 0$ and therefore A is injective.

"$\Longleftarrow$" Suppose $\beta > 0$. Then the restriction of B_1 to $V^\perp := \{x \in X : x \perp V\}$, namely

$$B: V^\perp \to Y^*, \quad x \mapsto b(x, \cdot),$$

is an isomorphism by Theorem 2.2, since (iv) and hence (a)-(b) hold with

$$\inf_{\substack{x \in V^\perp \\ \|x\|_X = 1}} \sup_{\substack{y \in Y \\ \|y\|_Y = 1}} b(x, y) = \beta > 0.$$

For given $(x, y) \in X \times Y$ with $\|x\|_X^2 + \|y\|_Y^2 = 1$, there exist $x_1 \in V$, $x_2 \in V^\perp$ with $x = x_1 + x_2$. Define $t := \|x_2\|_X$, $s := \|x_1\|_X$, $r := \|x\|_X$, and so $r^2 = s^2 + t^2 \le 1$ and $\|y\|_Y^2 = 1 - r^2$. By definition of α, β and the above inf-sup condition, there exist $u_1 \in V$, $u_2 \in V^\perp$ and $w \in Y$ with

$$\begin{aligned} a(x_1, u_1) &\ge \alpha \|x_1\|_X \|u_1\|_X, \\ b(x_2, w) &\ge \beta \|x_2\|_X \|w\|_Y, \\ b(u_2, y) &\ge \beta \|u_2\|_X \|y\|_Y. \end{aligned}$$

The lengths of u_1, u_2, w are arbitrary, and with $(\cdot)_+ := \max\{0, \cdot\}$ one possible choice reads

$$\begin{aligned} \|w\|_Y &:= \beta t \ge 0, \\ \|u_1\|_X &:= \big(\alpha s - t \|a\|\big)_+, \\ \|u_2\|_X &:= \big(\beta(1 - r^2)^{1/2} - (t + s) \|a\|\big)_+. \end{aligned}$$

With $u = u_1 + u_2$ this implies

$$\begin{aligned} L(x, y)(u, w) &= a(x_1 + x_2,\, u_1 + u_2) + b(u_1 + u_2,\, y) \\ &\qquad + b(x_1 + x_2, w) \\ &\ge \alpha \|x_1\|_X \|u_1\|_X - \|a\| \|x_2\|_X \|u_1 + u_2\|_X \\ &\qquad - \|a\| \|x_1\|_X \|u_2\|_X + \beta \|u_2\|_X \|y\|_Y \\ &\qquad + \beta \|x_2\|_X \|w\|_Y \\ &\ge \|u_1\|_X (\alpha s - \|a\| t) + \|w\|_Y \beta t \\ &\qquad + \|u_2\|_X (\beta(1 - r^2)^{1/2} - \|a\| t - \|a\| s) \\ &= \|u\|_X^2 + \|w\|_Y^2 \\ &= \big(\alpha s - \|a\| t\big)_+^2 \\ &\qquad + \big(\beta(1 - r^2)^{1/2} - (t + s) \|a\|\big)_+^2 + \beta^2 t^2 \\ &=: f(r, s, t). \end{aligned}$$

Since $f(r, s, t) = 0$ implies $s = 0 = t$ and $r = 1$, there holds

$$0 < \bar\beta^2 := \min\{f(r, s, t) \,:\, 0 \le r, s, t \le 1 \text{ with } r \le s + t\}.$$

The minimum $\bar\beta^2$ depends only on α, β, and $\|a\|$. Since $\|(u, w)\|_{X \times Y}^2 := \|u\|_X^2 + \|w\|_Y^2$, there holds

$$L(x, y)(u, w) \ge f(r, s, t) \ge \bar\beta \, \|(u, w)\|_{X \times Y}.$$

Therefore, L satisfies the inf-sup condition.

It remains to show that

$$\forall(u,w) \in (X \times Y) \setminus \{0\} \quad \exists(x,y) \in X \times Y,\ L(x,y)(u,w) \neq 0.$$

In fact, if $L(x,y)(u,w) = 0$ holds for all (x,y), then $b(u,\cdot) = 0$ and $a(\cdot,u) + b(\cdot,w) = 0$, hence $u \in V \subseteq X$.
Since a is non-degenerated and $u \in V$, the assumption $u \neq 0$ leads to some $z \in V$ with $a(z,u) \neq 0$ and $b(z,w) = 0$. This contradicts $a(z,u) + b(z,w) = 0$ and proves $u = 0$.

The assumption $w \neq 0$ and the Hahn-Banach theorem leads to some $f \in Y^*$ with $f(w) = 1$. Since the above operator B is an isomorphism, $x := B^{-1}(f) \in V^\perp$. This implies $b(x,w) = f(w) = 1$. But this contradicts $b(x,w) = 0$ and proves $w = 0$.

In conclusion, $(u,w) = 0$ and so L is non-degenerated. □

3 Discrete Mixed Formulations

The understanding of MFEM and its stability requires some mathematics, which will be introduced in this lecture. The instable methods sometimes even seem to work somehow in practice: *There are methods which work on Mondays but not on Tuesdays* is a saying due to Rolf Stenberg. However, a reliable computation is based on stable discretisations and therefore, has to satisfy the LBB conditions.

3.1 Abstract Framework

Throughout this lecture, let $a : X \times Y \to \mathbb{R}$ be a bounded bilinear form, and let $X_h \subseteq X$ and $Y_h \subseteq Y$ be closed (e.g., finite-dimensional) subspaces of the Hilbert spaces X and Y.

Definition 3.1 (LBB). The discrete bilinear form a is said to satisfy the (LBB) condition if

$$\alpha_h := \inf_{\substack{x_h \in X_h \\ \|x_h\|_X = 1}} \sup_{\substack{y_h \in Y_h \\ \|y_h\|_Y = 1}} a(x_h, y_h) \geq 0 \tag{9}$$

satisfies $\alpha_h > 0$ uniformly with respect to a family $(X_h, Y_h)_h$ of discrete spaces. LBB abbreviates Ladyzhenskaya, Babuška, Brezzi and is also called (discrete) inf-sup condition.

Remark 3.2. Any choice of some orthonormal basis $(\xi_1, \dots, \xi_m)$ in X_h and some orthonormal basis $(\eta_1, \dots, \eta_n)$ in Y_h leads to a coefficient matrix $A \in \mathbb{R}^{m\times n}$ via

$$A_{jk} := a(\xi_j, \eta_k) \quad \text{for } i = 1, \dots, m \text{ and } k = 1, \dots, n.$$

Then, α_h is the smallest singular value of A. This is seen from a singular value decomposition

$$A = Q\Sigma R$$

of A with a diagonal matrix $\Sigma \in \mathbb{R}^{m\times n}$ and orthogonal matrices $Q \in \mathbb{R}^{m\times m}$ and $R \in \mathbb{R}^{n\times n}$. In fact,

$$x_h = \sum_{j=1}^{m} \lambda_j \xi_j \neq 0 \quad \text{and} \quad y_h = \sum_{k=1}^{n} \mu_k \eta_k \neq 0$$

for real coefficients $\lambda_1, \dots, \lambda_m, \mu_1, \dots, \mu_n$ and (recall orthonormality of the basis)

$$\|x_h\|_X^2 = \sum_{j=1}^{m} \lambda_j^2 \quad \text{and} \quad \|y_h\|_Y^2 = \sum_{k=1}^{m} \mu_k^2.$$

Since the Euclidian norm $|\cdot|$ is invariant under orthonormal transformations, the vectors $\lambda = (\lambda_1, \dots, \lambda_m)^T$ and $\mu = (\mu_1, \dots, \mu_n)^T$ satisfy

$$\frac{a(x_h, y_h)}{\|x_h\|_X \, \|y_h\|_Y} = \frac{\lambda\, Q\Sigma R\, \mu}{|\lambda|\, |\mu|} = \frac{(Q^T\lambda)\, \Sigma\, (R\mu)}{|Q^T\lambda|\, |R\mu|}$$

and recall that Σ is diagonal with singular values $\sigma_1 \geq \sigma_2 \geq \ldots \geq \sigma_r \geq 0$ for $r = \min\{m, n\}$ in the diagonal, thus

$$\begin{aligned}
\alpha_n &= \inf_{\substack{\lambda \in \mathbb{R}^m \\ |\lambda| = 1}} \ \sup_{\substack{\mu \in \mathbb{R}^n \\ |\mu| = 1}} \sum_{j=1}^{m} \sum_{k=1}^{n} \lambda_j \, \Sigma_{jk} \, \mu_k \\
&= \inf_{\substack{\lambda \in \mathbb{R}^m \\ |\lambda| = 1}} \ \sup_{\substack{\mu \in \mathbb{R}^n \\ |\mu| = 1}} \sum_{\ell=1}^{r} \lambda_\ell \, \Sigma_{\ell\ell} \, \mu_\ell \\
&= \inf_{\substack{\lambda \in \mathbb{R}^m \\ |\lambda| = 1}} \ \sup_{\substack{\mu \in \mathbb{R}^n \\ |\mu| = 1}} \sum_{\ell=1}^{r} (\sigma_\ell \lambda_\ell) \, \mu_\ell.
\end{aligned}$$

Since the scalar product of $(\sigma_\ell \lambda_\ell)_{\ell=1,\dots,r}$ and $(\mu_1, \dots, \mu_r)$ is maximal if the two vectors are parallel,

$$\sup_{\substack{\mu \in \mathbb{R}^n \\ |\mu| = 1}} \sum_{\ell=1}^{r} (\sigma_\ell \lambda_\ell) \, \mu_\ell = \Big(\sum_{\ell=1}^{r} \sigma_\ell^2 \lambda_\ell^2 \Big)^{1/2}.$$

Hence, there holds

$$\alpha_h = \inf_{\substack{\lambda \in \mathbb{R}^m \\ |\lambda| = 1}} \left(\sum_{\ell=1}^{r} \sigma_\ell^2 \lambda_\ell^2 \right)^{1/2}.$$

This is zero if $n > r$, and equals the singular value σ_r if $n \leq r$.

It is emphasised here, that $m = n$ if A is regular and then α_h is the smallest (positive) singular value of A.

Definition 3.3 (Exact Problem). Given $f \in Y^*$, find $x \in X$ with

$$a(x, y) = f(y) \quad \text{for all } y \in Y. \tag{P}$$

Definition 3.4 (Discrete Problem). Given $f \in Y^*$, find $x_h \in X_h$ with

$$a(x_h, y_h) = f(y_h) \quad \text{for all } y_h \in Y_h. \tag{P_h}$$

Theorem 3.5 (Existence and Uniqueness).

(a) *The discrete problem (P_h) has a unique solution $x_h \in X_h$ for all $f \in Y^*$ if $\alpha_h > 0$ and a is non-degenerated in the sense of*

$$\forall y_h \in Y_h \; \exists \, x_h \in X_h, \; a(x_h, y_h) \neq 0. \tag{ND}$$

(Analogous results hold for the exact problem.)

(b) *Assume (P) to be uniquely solvable and let $x \in X$ be its unique solution. Furthermore, suppose $\alpha_h > 0$ in (9), (ND), and that x_h solves (P_h). Then*

$$\|x - x_h\|_X \leq (1 + \|a\|/\alpha_h) \inf_{v_h \in X_h} \|x - v_h\|_X \, . \tag{10}$$

Remark 3.6. The estimate (10) is called quasi-optimal, if $\alpha_h \geq \alpha > 0$ is independent of $\dim X_h$ for a family of discrete spaces $(X_h)_h$. It links the Galerkin error $\|x - x_h\|_X$ with the best-approximation error $\operatorname{dist}(x, X_h)$ by

$$\|x - x_h\| \approx \operatorname{dist}(x, X_h).$$

Proof of Theorem 3.5.

(a) This follows from Theorem 2.2 by replacing $X \times Y$ with $X_h \times Y_h$.

(b) For the exact solution x and a discrete solution x_h there holds the Galerkin orthogonality

$$a(x - x_h, y_h) = 0 \text{ for all } y_h \in Y_h.$$

Given $v_h \in X_h$ and $\varepsilon > 0$, (9) and $\alpha_h > 0$ imply that there exists $y_h \in Y_h$ with $\|y_h\|_Y = 1$ and

$$\begin{aligned}(\alpha_h - \varepsilon)\,\|x_h - v_h\|_X &\le a(x_h - v_h, y_h)\\ &= a(x - v_h, y_h)\\ &\le \|a\|\,\|x - v_h\|_X\,.\end{aligned}$$

Here, the second step is via the aforementioned Galerkin orthogonality and the last step is the boundedness of a. Since $\varepsilon > 0$ is arbitrarily small, this implies

$$\|x_h - v_h\|_X \le \frac{\|a\|}{\alpha_h}\,\|x - v_h\|_X\,.$$

This and a triangle inequality show the assertion:

$$\begin{aligned}\|x - x_h\|_X &\le \|x - v_h\|_X + \|v_h - x_h\|_X\\ &\le \left(1 + \frac{\|a\|}{\alpha_h}\right)\|x - v_h\|_X\,.\end{aligned}$$ □

3.2 Stable and Instable MFEM

This subsection considers one instable and one stable MFEM, i.e., violating and satisfying the (uniform) LBB condition, respectively.

Consider the following discrete problem (MFEM) for $\Omega \subset \mathbb{R}^n$ and finite dimensional subspaces $X_h \subset H(\mathrm{div}, \Omega)$ and $Y_h \subset L^2(\Omega)$ from Subsection 1.3. Compute $(p_h, u_h) \in X_h \times Y_h$ that satisfy

$$\begin{aligned}(p_h, q_h)_{L^2(\Omega)} + (u_h, \mathrm{div}\, q_h)_{L^2(\Omega)} &= 0 \quad \text{for all } q_h \in X_h,\\ (\mathrm{div}\, p_h, v_h)_{L^2(\Omega)} &= -(f, v_h)_{L^2(\Omega)} \quad \text{for all } v_h \in Y_h.\end{aligned} \tag{11}$$

Definition 3.7 (Triangulation). A regular triangulation $\mathcal{T}$ into triangles (for $n = 2$) is a set of closed triangles T of positive area $|T|$ such that any two distinct triangles T_1 and T_2 are either disjoint, $T_1 \cap T_2 = \emptyset$, or share exactly one vertex z, $T_1 \cap T_2 = \{z\}$, or have one edge $E = T_1 \cap T_2$ in common. The set of all edges is denoted by $\mathcal{E}$, the set of nodes is denoted by $\mathcal{N}$. Each edge is associated to a length $h_E := \mathrm{diam}(E)$ and a unit normal and unit tangential vector ν_E and τ_E. The words mesh and triangulation are used as synonyms of each other.

Definition 3.8 (FE Spaces). For a triangulation $\mathcal{T}$ of Ω and $k = 0, 1, 2, \ldots$ define

$$\begin{aligned} P_k(T) &:= \{\text{polynomials on } T \text{ with degree } \le k\},\\ P_k(\mathcal{T}) &:= \{f \in L^2(\Omega) \,:\, \forall T \in \mathcal{T}\ f|_T \in P_k(T)\},\\ \mathcal{S}^k(\mathcal{T}) &:= \{f \in P_k(\mathcal{T}) \,:\, f \text{ globally continuous in } \overline{\Omega}\},\\ \mathcal{S}^k_0(\mathcal{T}) &:= \{f \in \mathcal{S}^k(\mathcal{T}) \,:\, f = 0 \text{ on } \partial\Omega\}. \end{aligned}$$

Moreover, the nodal basis function $\phi_z \in \mathcal{S}^1(\mathcal{T})$ is defined by $\phi_z(z) = 1$ and $\phi_z(y) = 0$ for $z \in \mathcal{N}$ and all other nodes $y \in \mathcal{N} \setminus \{z\}$. One can prove that $(\varphi_z \,:\, z \in \mathcal{N})$ is indeed a basis of $\mathcal{S}^1(\mathcal{T})$ and $(\varphi_z \,:\, z \in \mathcal{N} \cap \Omega)$ is a basis of $\mathcal{S}^1_0(\mathcal{T})$.

Example 3.9 (Unstable P_1-P_0 MFEM). The 2D $P_1 - P_0$ MFEM is defined by the discrete spaces

$$X_h := (\mathcal{S}^1(\mathcal{T}))^2 \quad \text{and} \quad Y_h := P_0(\mathcal{T}).$$

As an example for instability of the linear system of equations underlying (P_h), consider the Poisson problem (4) on an L-shaped domain for $g \equiv 1$. A simple implementation that results after some uniform refinement in a singular matrix A of the linear system associated with (P_h) is depicted in Figure 1. The outcome is that the program stops with the error message "Warning: Matrix is singular to working precision." To illustrate this singularity, Figure 3.2 shows the smallest singular value λ of the energy matrix A with respect to the degrees of freedom. For $\max_{E \subset \mathcal{E}} h_E \to 0$ the singular value λ goes to 0. This yields instability as it can be observed in Figure 3.

Remark 3.10. Based on the Gauß divergence theorem, one can prove that any piecewise polynomial function $q_h \in P_k(\mathcal{T})^n \subset L^2(\Omega)^n$ belongs to $H(\operatorname{div}, \Omega)$ if and only if the jump of the normal component of q_h across E vanishes, i.e.,

$$[q_h]_E \cdot \nu_E = 0 \quad \text{on each interior edge } E.$$

This condition clearly indicates that the $P_1 - P_0$ MFEM is too restrictive by the request of global continuity.

Definition 3.11 (Raviart-Thomas FEM). For a regular triangulation $\mathcal{T}$ of $\Omega \subseteq \mathbb{R}^2$ into triangles set

$$RT_k(T) := \left\{ (x, y) \mapsto \begin{pmatrix} p_1(x, y) \\ p_2(x, y) \end{pmatrix} + p_3(x, y)\binom{x}{y} \,:\, p_1, p_2, p_3 \in P_k(T) \right\}$$

```
% Computation of local Stiffness Matrices
B = zeros(6,6,nrElems); C = zeros(1,6,nrElems);
genericMass = [ 2 1 1 0 0 0;
                1 2 1 0 0 0;
                1 1 2 0 0 0;
                0 0 0 2 1 1;
                0 0 0 1 2 1;
                0 0 0 1 1 2 ] /12;
genericC = [1 0 -1 0 1 -1];

for curElem = 1:nrElems
       area = area4e(curElem);
       B(:,:,curElem) = area*genericMass;
       C(:,:,curElem) = area*genericC;
end

% local DoF -> global DoF for matrix B
globalDoF = [n4e,nrNodes + n4e];
globalDoF_T = globalDoF';
I = repmat(globalDoF_T(:),1,size(globalDoF,2))';
J = repmat(globalDoF,1,size(globalDoF,2))';
B = sparse(I(:),J(:),B(:));

% local DoF -> global DoF for matrix C
I = repmat((1:nrElems),6,1);
J = [n4e,nrNodes + n4e]';
C = sparse(I(:),J(:),C(:),nrElems,2*nrNodes);

A = [B      C';
     C      sparse(nrElems,nrElems)];

% Assembling RHS
b = zeros(2*nrNodes + nrElems,1);
b(2*nrNodes+1:end) = -area4e .* ...
                       f(midPoint4e(:,1),midPoint4e(:,2),p);
% Solve
x = A\b;
```

Figure 1. MATLAB implementation of $P_1 - P_0$ MFEM: Not recommended in practice.

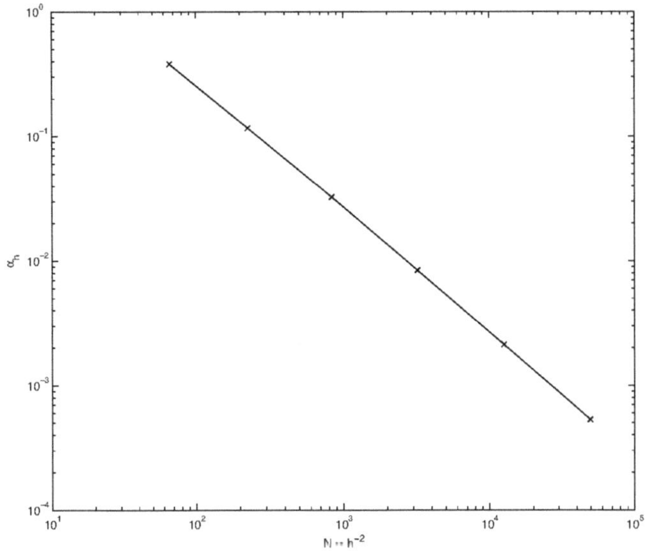

Figure 2. Smallest singular value α_h of the stiffness matrix of b from (12) for the instable $P_1 - P_0$ MFEM as a function of the degrees of freedom $N \propto h^{-2}$ in logarithmic scale.

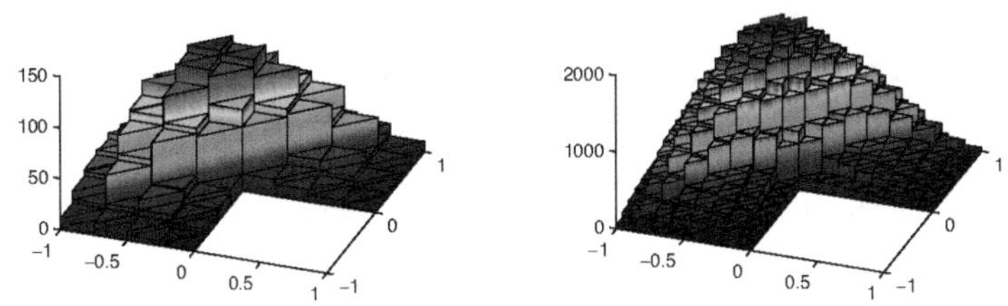

(a) Solution with 226 Degrees of Freedom

(b) Solution with 834 Degrees of Freedom

Figure 3. Illustration of the instability of $P_1 - P_0$ MFEM. Notice the different scaling in the left and the right picture, which indicates the instability.

for $k = 0, 1, \ldots$ and $T \in \mathcal{T}$. The Raviart-Thomas FE-Space RT_k of degree k is defined by

$$RT_k(\mathcal{T}) := \{p \in L^2(\Omega; \mathbb{R}^2) : \forall T \in \mathcal{T},\, p|_T \in RT_k(T)\},$$

$RT_k := RT_k(\mathcal{T}) \cap H(\operatorname{div}, \Omega)$. It is known that

$$RT_k = \{p \in RT_k(\mathcal{T}) : \forall E \in \mathcal{E}_\Omega,\, [p]_E \cdot \nu_E = 0 \text{ on } E\}.$$

Theorem 3.12. *The bilinear form*

$$a : \begin{cases} H(\operatorname{div}, \Omega) \times L^2(\Omega) \to \mathbb{R}, \\ (p, u) \mapsto \int_\Omega u \operatorname{div} p \, dx \end{cases} \tag{12}$$

satisfies the (LBB) condition for $RT_k(\mathcal{T}) \times P_k(\mathcal{T})$. Therefore, the discrete solutions satisfy quasi-optimal convergence in the sense that

$$\begin{aligned} &\|p - p_h\|_{H(\operatorname{div},\Omega)} + \|u - u_h\|_{L^2(\Omega)} \\ &\qquad \lesssim \inf_{q_h \in RT_k(\mathcal{T})} \|p - q_h\|_{H(\operatorname{div},\Omega)} + \inf_{v_h \in P_k(\mathcal{T})} \|u - v_h\|_{L^2(\Omega)}, \end{aligned}$$

where (p, u) denotes the exact solution.

Proof. The proof is based on Fortin's lemma and the property $\operatorname{div} RT_0(\mathcal{T}) = P_0(\mathcal{T})$. Details can be found in the literature Brezzi and Fortin (1991). □

Remark 3.13. The list of $2D$ examples for the Stokes problem includes conforming finite elements such as the MINI element, the P_2-P_0 finite element, and the Taylor-Hood element Brezzi and Fortin (1991) and the non-conforming finite element due to Kouhia and Stenberg Kouhia and Stenberg (1995).
There are also finite element methods for the unsymmetric formulation Brezzi and Fortin (1991) such as the popular Crouzeix-Raviart finite element. Since the arguments of this lecture work verbatim (if not simpler), we omit details and refer to Dari et al. (1995); Verfürth (1989).

Remark 3.14. The MFEM for elasticity essentially started with PEERS (see Subsection 8.2 on page 46) and has been developed for symmetric stress approximation very recently by Arnold and Winther Arnold and Winther (2002).

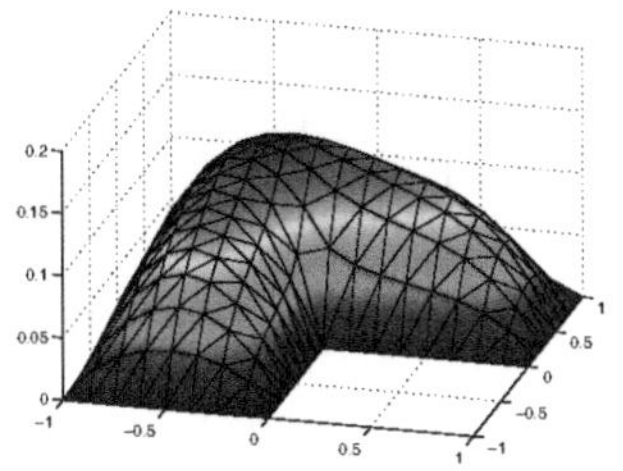

(a) P_1 approximation u_h from conforming FEM

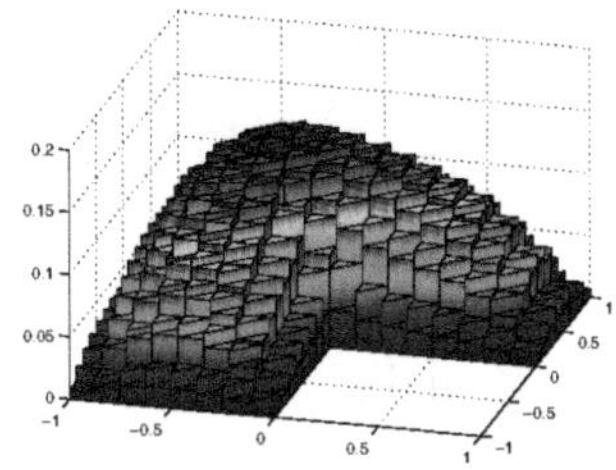

(b) P_0 approximation u_h from RT_0 MFEM

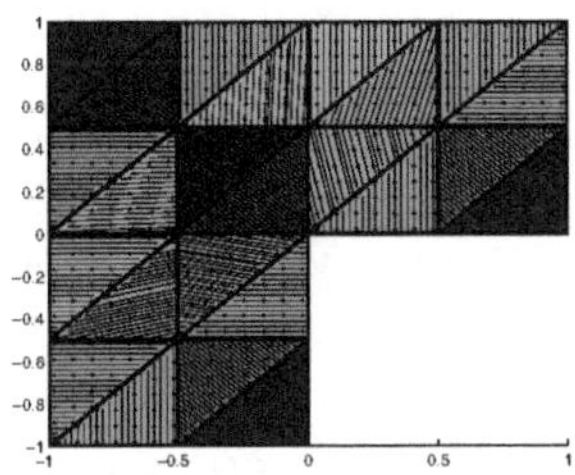

(c) P_0 approximation ∇u_h from conforming FEM

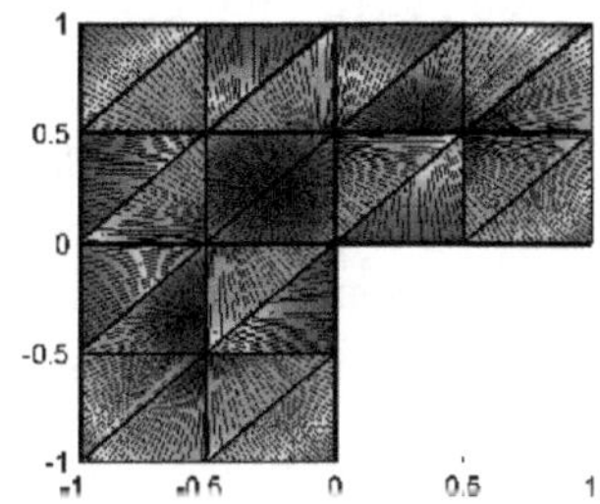

(d) RT_0 approximation p_h from RT_0 MFEM

Figure 4. Visualisation of u_h and p_h from P_1 FEM and RT_0 MFEM for the Poisson problem $-\Delta u = 1$ on the depicted L-shaped domain Ω with $u = 0$ on $\partial\Omega$ and two meshes with 833 nodes (top) and 21 nodes (bottom).

3.3 Numerical Realisation of RT_0 MFEM

This subsection is devoted to the edge-basis functions Ψ_E for the lowest order Raviart-Thomas finite elements employed in the MATLAB program EBmfem that realises the resulting linear system of equations of the typical format

$$A = \begin{pmatrix} B & C \\ C^T & 0 \end{pmatrix} \begin{pmatrix} x_\Psi \\ x_u \end{pmatrix} = \begin{pmatrix} b_D \\ b_f \end{pmatrix}. \tag{13}$$

Definition 3.15 (Local Definition of Ψ_E). Let E_1, E_2, E_3 be the edges of a triangle T opposite its vertices P_1, P_2, P_3, respectively, and let ν_{E_j} denote the unit normal vector of E_j chosen with a global fixed orientation while ν_j

denotes the outer normal unit of T along E_j. Define

$$\Psi_{E_j}(x) = \sigma_j \frac{|E_j|}{2|T|}(x - P_j) \qquad \text{for } j = 1,2,3 \text{ and } x \in T,$$

where $\sigma_j = \nu_j \cdot \nu_{E_j}$ is $+1$ if ν_{E_j} points outwards and else -1; $|E_j|$ is the length of E_j, and $|T|$ is the area of T,

$$2|T| = \det(P_2 - P_1, P_3 - P_1) = \det \begin{pmatrix} P_1 & P_2 & P_3 \\ 1 & 1 & 1 \end{pmatrix}$$

(with the 3×3 matrix that consists of the 2×3 matrix of the three vectors $P_1, P_2, P_3 \in \mathbb{R}^2$ plus three ones in the last row).

Definition 3.16 (Notation for Elements that Share an Edge E)**.** The lower index $\pm$ in the subsequent notation refers to two cases namely, the case $(\,)_+$ and the case $(\,)_-$ separately. Let $T_\pm = \operatorname{conv}\{E \cup \{P_\pm\}\}$ for the vertex $P_\pm$ opposite to E of $T_\pm$ such that the edge $E = \operatorname{conv}\{A, B\}$ orients from A to B as depicted in Figure 5. Then ν_E points outward from T_+ into T_- with a positive sign. If E is an exterior edge $E \in \mathcal{E} \setminus \mathcal{E}_\Omega$, $\mathcal{E}_\Omega := \{E \in \mathcal{E} : \exists T_+, T_- \in \mathcal{T} \;\; E = T_+ \cap T_-\}$, then $\nu = \nu_E$ is the exterior normal and $E \subset \partial T_+$ defines T_+ (and T_- is undefined). Moreover, the unit tangent vector τ_E points in the direction of E from A to B.

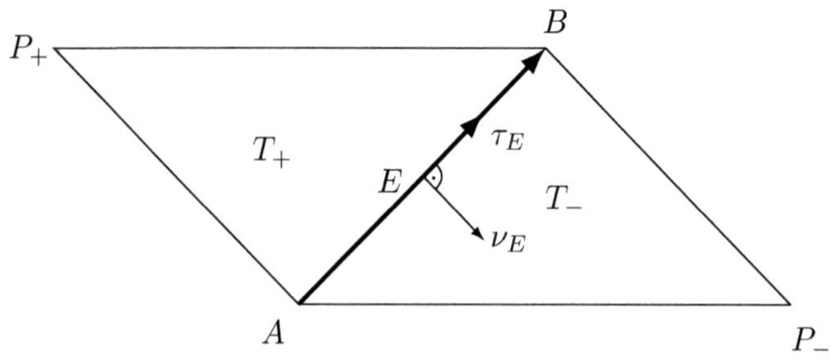

Figure 5. Two neighbouring triangles T_+ and T_- that share the edge $E = \partial T_+ \cap \partial T_-$ with initial node A and end node B and unit normal ν_E. The orientation of ν_E equals the outer normal of T_+ (and hence points into T_-).

Note that the normal and tangential directions change for reverse orientation of $E = \operatorname{conv}\{A, B\}$.

Definition 3.17 (Global Definition of Ψ_E). Given an edge $E \in \mathcal{E}$ there are either two elements T_+ and T_- in $\mathcal{T}$ with the joint edge $E = \partial T_+ \cap \partial T_-$ or there is exactly one element T_+ in $\mathcal{T}$ with $E \subset \partial T_+$. Then if $T_\pm = \operatorname{conv}\{E \cup \{P_\pm\}\}$ for the vertex $P_\pm$ opposite to E of $T_\pm$ set

$$\Psi_E(x) := \begin{cases} \pm\frac{|E|}{2|T_\pm|}(x - P_\pm) & \text{for } x \in T_\pm, \\ 0 & \text{elsewhere.} \end{cases} \tag{14}$$

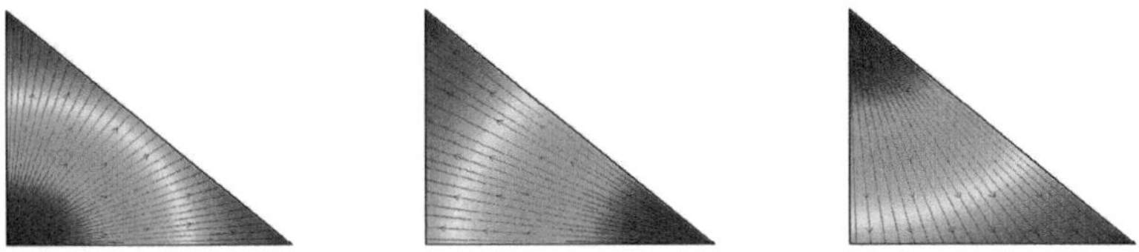

(a) Ψ_{E_1} (b) Ψ_{E_2} (c) Ψ_{E_3}

Figure 6. Vector fields of the Raviart-Thomas basis functions, the color visualises the norm.

Lemma 3.18. *There hold*

(a) $\Psi_E \cdot \nu_F = \begin{cases} 0 & \textit{along } F \in \mathcal{E} \setminus \{E\}, \\ 1 & \textit{along } F = E; \end{cases}$

(b) $\Psi_E \in H(\operatorname{div}, \Omega)$*;*

(c) $(\Psi_E : E \in \mathcal{E})$ *is a basis of* $RT_0(\mathcal{T})$*;*

(d) $\operatorname{div} \Psi_E = \begin{cases} \pm\frac{|E|}{|T_\pm|} & \textit{on } T_\pm, \\ 0 & \textit{elsewhere.} \end{cases}$

Proof.

(a) & (b) Consider $T_\pm = \operatorname{conv}\{P_\pm, A, B\}$ with $E = T_+ \cap T_-$ and denote $\omega_E := T_+ \cup T_-$. Let $F \in \mathcal{E} \setminus \{E\}$ be an edge different from E. Clearly, $\Psi_E \cdot \nu_F$ vanishes for $F \not\subset \partial\omega_E$. For $x \in F \subset \partial\omega_E$ the vector $x - P_\pm$ is tangential to $\partial\omega_E$ and hence $\Psi_E(x) \cdot \nu_F = 0$ as well. For $x \in E = F$, $(x - P_\pm) \cdot \nu_E$ is the height of the triangle $T_\pm$. Hence it is constant and equals $2|T_\pm|/|E|$. The factor in Definition 3.17 then yields $\Psi_E(x) \cdot \nu_E = 1$. As this is independent of the choice of T_+ or T_-, we get $[\Psi_E]_F \cdot \nu_F = 0$ for all interior edges F in $\mathcal{E}$. Remark 3.10 then gives $\Psi_E \in H(\operatorname{div}, \Omega)$.

(c) The functions $(\Psi_E : E \in \mathcal{E})$ are (uniquely) determined by $\Psi_E \in RT_0(\mathcal{T})$ and $\Psi_E \cdot \nu_F = 1$ for $E = F \in \mathcal{E}$ and $\Psi_E \cdot \nu_F = 0$ for $E \neq F$ in $\mathcal{E}$. Given any $q_h \in RT_0(\mathcal{T})$ notice that $q_h \cdot \nu_E$ is constant on $E \in \mathcal{E}$ and define

$$p_h := q_h - \sum_{E \in \mathcal{E}} (q_h \cdot \nu_E) \Psi_E \in RT_0(\mathcal{T}).$$

Then, (a) implies $p_h \cdot \nu_E = 0$ for all $E \in \mathcal{E}$. On $T \in \mathcal{T}$ with edges E_1, E_2, E_3, there holds

$$p_h(P_j) \cdot \nu_{E_k} = 0 \qquad \text{for } k = \{1, 2, 3\} \setminus \{j\}$$

at the vertex P_j opposite to E_j. Since $p_h|_T$ is affine, this proves $p_h \equiv 0$ on $T \in \mathcal{T}$. Consequently, $RT_0(\mathcal{T}) \subseteq \operatorname{span}\{\Psi_E : E \in \mathcal{E}\}$. It remains to prove that $\{\Psi_E : E \in \mathcal{E}\}$ are linearly independent. Given coefficients ξ_E for any $E \in \mathcal{E}$ with

$$p_h \equiv 0 \equiv \sum_{E \in \mathcal{E}} \xi_E \, \Psi_E \quad \text{in } \Omega,$$

(a) implies $0 \equiv p_h|_F \cdot \nu_F = \xi_F$ for all $F \in \mathcal{E}$.

(d) This follows from (14). □

Definition 3.19 (Local Stiffness Matrices)**.** Adopt notation from above for a triangle T with edges E_1, E_2, E_3 with (local) number $j = 1, 2, 3$ and abbreviate $\Psi_j := \Psi_{E_j}$. Then,the local stiffness matrices $B_T, C_T \in \mathbb{R}^{3\times 3}$ read

$$(B_T)_{jk} := \int_T \Psi_j \cdot \Psi_k \, dx \qquad \text{for } j, k = 1, 2, 3; \tag{15}$$

$$C_T := \operatorname{diag}\left(\int_T \operatorname{div} \Psi_1 \, dx, \int_T \operatorname{div} \Psi_2 \, dx, \int_T \operatorname{div} \Psi_3 \, dx \right). \tag{16}$$

Some direct calculations allow for an explicit representation of B_j.

Lemma 3.20. *Given* (15)-(16) *and the matrices* $M \in \mathbb{R}^{6\times 6}$ *and* $N \in \mathbb{R}^{6\times 3}$

$$M := \begin{pmatrix} 2 & 0 & 1 & 0 & 1 & 0 \\ 0 & 2 & 0 & 1 & 0 & 1 \\ 1 & 0 & 2 & 0 & 1 & 0 \\ 0 & 1 & 0 & 2 & 0 & 1 \\ 1 & 0 & 1 & 0 & 2 & 0 \\ 0 & 1 & 0 & 1 & 0 & 2 \end{pmatrix} \qquad N := \begin{pmatrix} 0 & P_1 - P_2 & P_1 - P_3 \\ P_2 - P_1 & 0 & P_2 - P_3 \\ P_3 - P_1 & P_3 - P_2 & 0 \end{pmatrix},$$

there holds

$$B_T = \frac{1}{48|T|} C_T^T N^T M N C_T.$$

Set $K = \mathrm{card}(\mathcal{E})$ and $L = \mathrm{card}(\mathcal{T})$. The global stiffness matrix A, namely the coefficient matrix in (13), consists of matrices $B \in \mathbb{R}^{K\times K}$ and $C \in \mathbb{R}^{K\times L}$ computed from the local stiffness matrices B_T and C_T. For a given element $T \in \mathcal{T}$, $\mathcal{E}|_T = \{E \in \mathcal{E} : E \subset \partial T\}$ denotes the set of its edges and ν_T its outer unit normal vector. Let m_k be the global edge number of $E_k \in \mathcal{E}|_T$ for $k = 1, 2, 3$ and ν_{E_k} be the unit normal vector of E_k. In a formal way, the assembling

$$B = \sum_{T\in\mathcal{T}} B(T) \quad \text{and} \quad C = \sum_{T\in\mathcal{T}} C(T)$$

requires the concept of the $K \times K$ matrix $B(T)$ defined by

$$B(T)\begin{pmatrix} m_1 & m_2 & m_3 \\ m_1 & m_2 & m_3 \end{pmatrix} = \mathrm{diag}\left(\sigma_{E_{m_1}}, \sigma_{E_{m_2}}, \sigma_{E_{m_3}}\right) B_T \,\mathrm{diag}\left(\sigma_{E_{m_1}}, \sigma_{E_{m_2}}, \sigma_{E_{m_3}}\right)$$

in the components m_1, m_2, m_3 for signs $\sigma_{E_{m_j}} = \nu_{E_{m_j}} \cdot \nu_T = \pm 1$, $j = 1, 2, 3$, and zero elsewhere; $C(T)$ is treated similarly.

3.4 MATLAB Realisation

The MATLAB realisation for the computation of the local stiffness matrix B_T in EBmfem via Lemma 3.20 reads

```
function B=stimaB(coord);
N=coord(:)*ones(1,3)-repmat(coord,3,1);
C=diag([norm(N([5,6],2)),norm(N([1,2],3)), ...
        norm(N([1,2],2))]);
M=spdiags([ones(6,1),ones(6,1),2*ones(6,1), ...
           ones(6,1),ones(6,1)],...
          [-4,-2,0,2,4],6,6);
B=C*N'*M*N*C/(24*det([1,1,1;coord]));
```

The MATLAB routines for the assembling of the global stiffness matrix B and matrix C read

```
B=sparse(noedges,noedges);
C=sparse(noedges,size(element,1));
for j=1:size(element,1)
  coord=coordinate(element(j,:),:)';
```

```
    I=diag(nodes2edge(element(j,[2 3 1]), ...
                      element(j,[3 1 2])));
    signum=ones(1,3);
    signum(find(j==edge2element(I,4)))=-1;
    n=coord(:,[3 1 2])-coord(:,[2 3 1]);
    B(I,I)=B(I,I)+diag(signum)*stimaB(coord)* ...
                   diag(signum);
    C(I,j)=diag(signum)*[norm(n(:,1)) ...
              norm(n(:,2)) norm(n(:,3))]';
end
```

and the global matrix A from (13) is generated by

```
A=sparse(noedges+size(element,1), ...
         noedges+size(element,1));
A=[B,C;C',sparse(size(C,2),size(C,2))];
```

3.5 Adaptive Mixed FEM (AMFEM)

It is the purpose of this subsection to outline an adaptive algorithm, AMFEM, and to illustrate its superiority over uniform mesh refinement in one numerical benchmark example. The theoretical background will be introduced in the following chapters.

The successive mesh refinement consists of loops of the subsequent form

$$\text{SOLVE} \Rightarrow \text{ESTIMATE} \Rightarrow \text{MARK} \Rightarrow \text{REFINE}.$$

Input: Coarse shape-regular triangulation $\mathcal{T}_0$.

For $\ell = 0, 1, 2, \ldots$ until termination do (a)-(d)

(a) (SOLVE) Compute (p_ℓ, u_ℓ) from (11) and $RT_k \times P_k$ MFEM.

(b) (ESTIMATE) Compute $\eta_\ell^2 := \sum_{E \in \mathcal{E}_\ell} \left(\eta_E^2 + \operatorname{osc}_E^2\right)$, with

$$\eta_E^2 := h_E \| [p_\ell]_E \|_{L^2(E)}^2 \quad \text{and} \quad \operatorname{osc}_E^2 := h_E^2 \| f - f_{\omega_E} \|_{L^2(\omega_E)}^2,$$

where $[p_\ell] := p_\ell|_{T_+} - p_\ell|_{T_-}$ and $f_{\omega_E} := |\omega_E|^{-1} \int_{\omega_E} f(x)\, dx$ on the patch $\omega_E := T_+ \cup T_-$, $T_\pm \in \mathcal{T}_\ell$, of the edge $E := \partial T_+ \cap \partial T_-$ as in Figure 5.

Check termination criterion based on $\|p - p_\ell\|_{L^2(\Omega)} \lesssim \eta_\ell$.

(c) (MARK) Find $\mathcal{M}_\ell \subset \mathcal{E}_\ell$ such that

$$\frac{1}{2}\eta_\ell^2 \leq \sum_{E \in \mathcal{M}_\ell} (\operatorname{osc}_E^2 + \eta_E^2).$$

(d) (REFINE) Generate shape-regular refinement $\mathcal{T}_{\ell+1}$ of $\mathcal{T}_\ell$ (no hanging nodes) where, at least, each edge $E \in \mathcal{M}_\ell$ is bisected based on the refinement rules of Figure 8.

Output: A (finite) sequence of MFEM approximations (p_ℓ, u_ℓ), triangulations $\mathcal{T}_\ell$, and error estimators η_ℓ.

The subsequent figures display the resulting sequence with some global refinement plus some balanced local refinement near the re-entering corner of the L-shape domain Ω. The convergence history in Figure 10 clearly displays the optimal linear convergence of the AMFEM compared to the reduced one for a uniform MFEM.

4 Second Analysis for Mixed Formulations

This lecture is devoted to some analytical tools required in the unified a posteriori error analysis below. The first is a Helmholtz decomposition for the bounded $2D$ Lipschitz domain Ω with boundary $\partial\Omega$ decomposed into connected closed subarcs $\Gamma_0, \ldots, \Gamma_J$.

Theorem 4.1. *Suppose $\Omega \subset \mathbb{R}^2$ is connected, $\partial\Omega = \Gamma_0 \cup \cdots \cup \Gamma_J$ and Γ_0 is the boundary of the unbounded compound of $\mathbb{R}^2 \setminus \Omega$. A function $q \in L^2(\Omega)^2$ satisfies*

$$\operatorname{div} q = 0 \quad \textit{and} \quad \int_{\Gamma_j} q \cdot \nu \, ds = 0 \quad \textit{for all } j = 0, ..., J$$

if and only if there exists a stream function $\phi \in H^1(\Omega)$ such that $q = \operatorname{Curl} \phi$. Moreover, if Ω is simply-connected, and Γ is $C^{1,1}$ or Ω is convex, then ϕ can be chosen such that

$$\|\nabla\phi\|_{L^2(\Omega)} = \|q\|_{L^2(\Omega)}. \tag{17}$$

Proof. The idea is to extend q to the whole plane such that it remains divergence-free. Then, it will be easy to construct its stream function ϕ by means of the Fourier transformation. For details see Girault and Raviart (1986). The converse implication follows from direct calculations. □

Remark 4.2. Theorem 4.1 holds essentially in $\mathbb{R}^3$ as well, with (17) replaced by $\|\nabla\phi\|_{L^2(\Omega)} \lesssim \|q\|_{L^2(\Omega)}$.

Theorem 4.3. *Given any $p \in L^2(\Omega; \mathbb{R}^2)$ for some simply-connected $\Omega \subset \mathbb{R}^2$, $J = 0$ in Theorem 4.1, there exist $\alpha \in H^1(\Omega)$ and $\beta \in H^1(\Omega)$ such that*

$$p = \nabla\alpha - \operatorname{Curl}\beta.$$

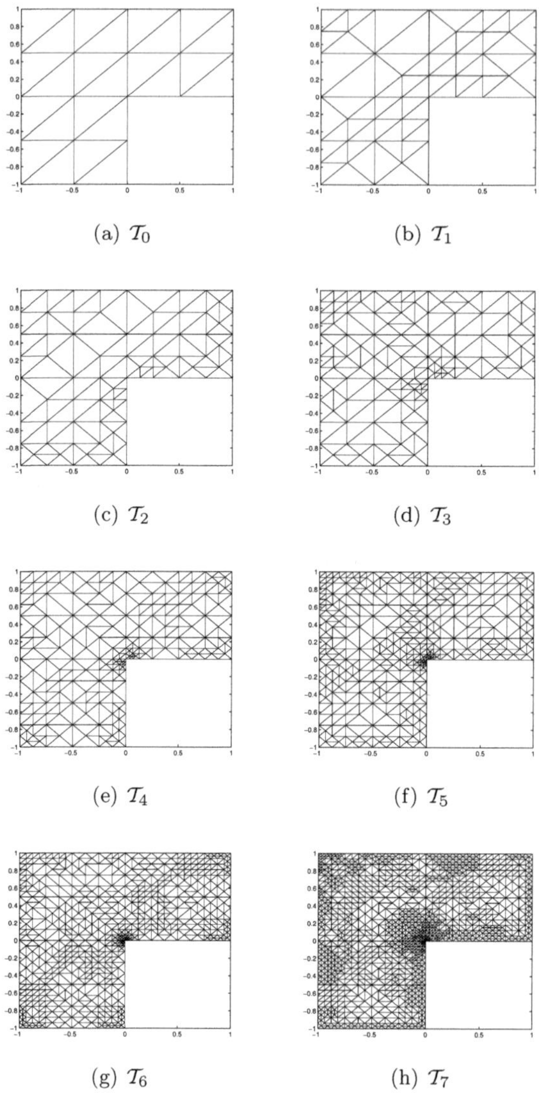

(a) $\mathcal{T}_0$ (b) $\mathcal{T}_1$

(c) $\mathcal{T}_2$ (d) $\mathcal{T}_3$

(e) $\mathcal{T}_4$ (f) $\mathcal{T}_5$

(g) $\mathcal{T}_6$ (h) $\mathcal{T}_7$

Figure 7. Triangulations $\mathcal{T}_0, \mathcal{T}_1, \ldots, \mathcal{T}_7$ generated by AMFEM with visible balance of local mesh-refinement towards the origin, the re-entering corner of the L-shape domain Ω, and global refinement everywhere else.

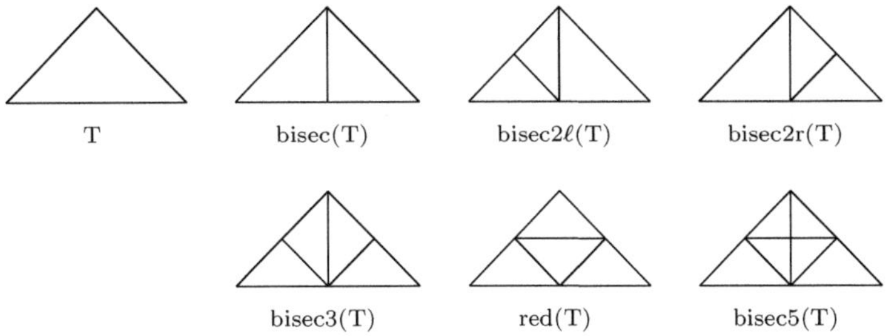

Figure 8. Possible refinements of one triangle T in the step REFINE up to rotation of the base-line, which is chosen for maintaining shape regularity.

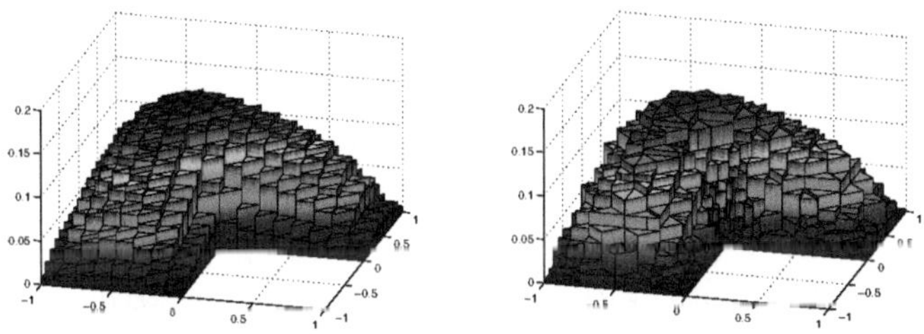

(a) uniform (b) adaptive

Figure 9. Visualisation of piecewise constant u_ℓ from RT_0 MFEM and RT_0 AMFEM for the Poisson problem $-\Delta u = 1$ on the depicted L-shaped domain Ω with $u = 0$ on $\partial\Omega$.

The functions α and β can be chosen such that

$$\alpha|_{\partial\Omega} = 0 \quad \textit{and} \quad \int_\Omega \beta\, dx = 0$$

denoted by $\alpha \in H_0^1(\Omega)$ and $\beta \in H^1(\Omega)/\mathbb{R}$.

Proof. Choose $\alpha \in H_0^1(\Omega)$ such that for all $v \in H_0^1(\Omega)$ there holds

$$\int_\Omega (\nabla\alpha - p) \cdot \nabla v\, dx = 0.$$

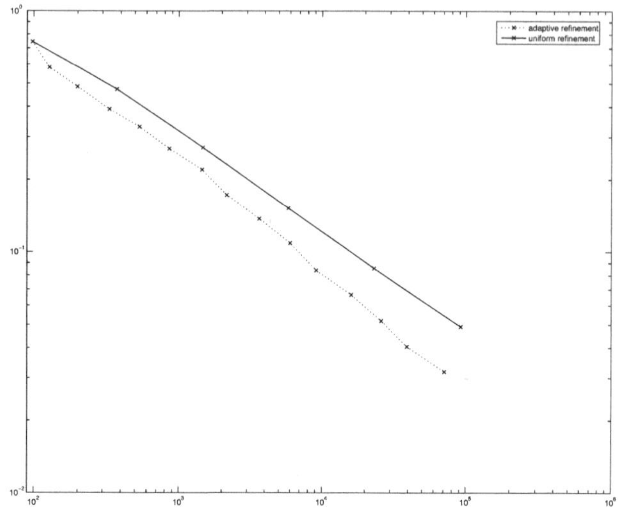

Figure 10. Convergence history: Estimated error $\|p-p_h\|_{L^2(\Omega)}$ of AMFEM (dashed line) and MFEM (solid line) vs number of degrees of freedom in logarithmic scale.

Then, $q := \nabla\alpha - p$ is divergence-free, $\operatorname{div} q = 0$ in Ω. The Gauss divergence theorem proves

$$\int_{\partial\Omega} q \cdot \nu \, ds = 0.$$

According to Theorem 4.1 (with $J = 0$), we can find $\beta \in H^1(\Omega)$ such that $q = \operatorname{Curl} \beta$. $\square$

Notice that, for all $\varphi \in H_0^1(\Omega)$ and all $\psi \in H^1(\Omega)$, there holds the orthogonality

$$\int_{\Omega} \nabla\varphi \cdot \operatorname{Curl} \psi \, dx = 0,$$

in particular, $\int_\Omega \nabla\alpha \cdot \operatorname{Curl} \beta \, dx = 0$.

The second set of analytical tools concerns a primal formulation of the mixed type.

Example 4.4. (Poisson Problem). The flux $p := \nabla u \in L^2(\Omega)^2$ and $u \in H_0^1(\Omega)$ of the Poisson problem (4) solve the problem

$$\begin{aligned} L(p,u)(q,v) := a(p,q) + b(p,v) + b(q,u) \overset{!}{=} -\int_\Omega gv\,dx \\ \text{for all } (q,v) \in X = L^2(\Omega)^2 \times H_0^1(\Omega) \end{aligned} \tag{18}$$

of the form considered in Subsection 1.3 with

$$a(p,q) := \int_\Omega p\cdot q\,dx \quad \text{and} \quad b(p,v) := -\int_\Omega p\cdot\nabla v\,dx.$$

The operator $L : X \to X^*$ defined in $(18)_\text{a}$ is bounded, linear and bijective. A direct proof of an inf-sup condition for L follows for any $(p,u) \in L^2(\Omega)^2 \times H_0^1(\Omega)$ from $(q,v) := (p - \nabla u; -2u) \in L^2(\Omega)^2 \times H_0^1(\Omega)$ and

$$\begin{aligned} 1/6\,\|(p,u)\|_X\|(q,v)\|_X &\le 1/6\,(\|p\|_{L^2(\Omega)} + \|u\|_{H_0^1(\Omega)})(\|p\|_{L^2(\Omega)} + 3\|u\|_{H_0^1(\Omega)}) \\ &\le \|p\|^2_{L^2(\Omega)} + \|u\|^2_{H_0^1(\Omega)} = L(p,u)(q,v). \end{aligned}$$

The (generalised) Lax-Milgram lemma then yields bijectivity of L.

Example 4.5. (Stokes Problem). Let $\operatorname{dev}\sigma := \sigma - 1/2\operatorname{tr}(\sigma)$ and $\operatorname{tr}(\sigma) := \sigma_{1,1} + \sigma_{2,2}$ denote the deviatoric part and the trace of σ respectively. Given a positive viscosity μ, define

$$\begin{aligned} a(\sigma,\tau) &:= \int_\Omega 1/(2\mu)\operatorname{dev}\sigma : \operatorname{dev}\tau\,dx \quad \text{for } \sigma,\tau \in L^2(\Omega,\mathbb{R}^{2\times 2}_{\text{sym}}), \\ b(\sigma;v) &:= -\int_\Omega \sigma : \varepsilon(v)\,dx \quad \text{for } (\sigma,v) \in L^2(\Omega;\mathbb{R}^{2\times 2}_{\text{sym}}) \times H_0^1(\Omega). \end{aligned} \tag{19}$$

Let $L^2(\Omega;\mathbb{R}^{2\times 2}_{\text{sym}})/\mathbb{R} := \{\sigma \in L^2(\Omega;\mathbb{R}^{2\times 2}) : \int_\Omega \operatorname{tr}\sigma\,dx = 0\}$ and $X := \big(L^2(\Omega;\mathbb{R}^{2\times 2}_{\text{sym}})/\mathbb{R}\big) \times H_0^1(\Omega)$. The operator $L : X \to X^*$, defined for (σ,u), $(\tau,v) \in X$ by

$$L(\sigma,u)(\tau,v) := a(\sigma,\tau) + b(\sigma,v) + b(\tau,u), \tag{20}$$

is linear, bounded and bijective. (This result holds for 2D and 3D.)

Example 4.6. (Linear Elasticity Problem). Given positive material (Lamé) parameters λ and μ, let

$$a(\sigma,\tau) := \int_\Omega (\mathbb{C}^{-1}\sigma) : \tau\,dx \quad \text{for } \sigma,\tau \in L^2(\Omega;\mathbb{R}^{2\times 2}_{\text{sym}})$$

replace $(19)_a$ and adopt $b(\sigma; v)$ from $(19)_b$. Then, the operator $L : X \to X^*$ from (20) is linear, bounded, and bijective and the operator norms of L and L^{-1} are λ-independent. A proof is included in Braess et al. (2004). The operator L represents the weak form of the Hellinger-Reissner principle in mechanics.

5 Residual-Based AFEM Error Control

This lecture is devoted to an a posteriori MFEM error control for the Poisson Problem.

Recall the Raviart-Thomas MFEM space and the piecewise constant space

$$\begin{aligned} RT_0 &:= \{q_h \in H(\mathrm{div}, \Omega) \,:\, \forall T \in \mathcal{T}\, \exists a \in \mathbb{R}^2\, \exists b \in \mathbb{R}\, \forall x \in T, \\ &\qquad q_h(x) = a + b\,x\}, \\ P_0(\mathcal{T}) &:= \{v_h \in L^\infty(\Omega) \,:\, \forall T \in \mathcal{T}\, \exists a \in \mathbb{R}\, \forall x \in T,\, v_h(x) = a\}. \end{aligned}$$

Recall

$$\eta := \left(\sum_{E \in \mathcal{E}} h_E \|[p_h]_E\|^2_{L^2(E)} \right)^{1/2}.$$

Definition 5.1. (Element Oscillations) Given a triangulation $\mathcal{T}$ and a function $f \in L^2(\Omega)$ with piecewise constant integral mean $f_\mathcal{T} \in P_0(\mathcal{T})$,

$$f_\mathcal{T}|_T := f_T := \fint_T f(x)\, dx := |T|^{-1} \int_T f(x)\, dx \quad \text{on any } T \in \mathcal{T},$$

define

$$\mathrm{osc}(f, \mathcal{T}) := \left(\sum_{T \in \mathcal{T}} h_T^2 \, \|f - f_\mathcal{T}\|^2_{L^2(T)} \right)^{1/2}.$$

Theorem 5.2 (Reliability and Efficiency)**.** *Let $p = \nabla u$ denote the exact flux in Example 4.4 and let p_h denote the discrete RT_0 MFEM solution. Then, there holds*

$$\eta \lesssim \|p - p_h\|_{L^2(\Omega)} \lesssim \eta + \mathrm{osc}(g, \mathcal{T}).$$

Remark 5.3. The reliability proof utilises a variable β_h, which is based on the commutative diagram

$$\begin{array}{ccc} H(\mathrm{div}, \Omega) & \xrightarrow{\mathrm{div}} & L^2(\Omega) \\ \Pi_h \downarrow & & \downarrow \Pi_{L_h} \\ M_h & \xrightarrow{\mathrm{div}} & L_h \end{array}$$

with some projection operators Π_h and Π_{L_h}. The same construction is possible for BDFM and MFEMs as well Carstensen (1999) as depicted in Table 1.

Element	$M_k(T)$	$D_k(T)$
RT	$P_k^2 + x \cdot P_k$	P_k
BDM	P_{k+1}^2	P_k
BDFM	$\{q \in P_{k+1}^2 : (q \cdot n)\|_{\partial T} \in \mathcal{R}_k(\partial T)\}$	P_k

Table 1. Standard 2D Mixed FEMs allowed in Theorem 6.2. Here, P_k denotes polynomials of total degree at most $k = 0, 1, 2, \ldots$ and $\mathcal{R}_k(\partial T)$ denotes (not necessarily continuous) functions on ∂T which equal a polynomial of degree at most k on each edge; see Carstensen (1999) for details and precise results.

Remark 5.4. The a posteriori error control of Theorem 5.2 was established independently in Alonso (1996) and Carstensen (1997).

Remark 5.5. The papers Braess et al. (2004); Carstensen and Dolzmann (1998); Carstensen (2005); Carstensen et al. (2000); Carstensen and Funken (2001b,a); Dari et al. (1995); Verfürth (1989) investigate the a posteriori error control for the Stokes and the elasticity problem.

The remaining part of this lecture is devoted to an outline of the proof of Theorem 5.2.

Given the exact and discrete solutions (u, p) and (u_h, p_h), the point of departure is the Helmholtz decomposition of the flux error $p - p_h$, namely

$$p - p_h = \nabla\alpha - \operatorname{Curl}\beta$$

for $\alpha \in H_0^1(\Omega)$ and $\beta \in H^1(\Omega)/\mathbb{R}$. Then

$$\|p - p_h\|_{L^2(\Omega)}^2 = \|\nabla\alpha\|_{L^2(\Omega)}^2 + \|\operatorname{Curl}\beta\|_{L^2(\Omega)}^2 .$$

The first key assertion reads

$$\|\nabla\alpha\|_{L^2(\Omega)} \lesssim \operatorname{osc}(g, \mathcal{T}). \tag{21}$$

By orthogonality of $\nabla\alpha$ and $\operatorname{Curl}\beta$ and an integration by parts, there holds

$$\begin{aligned} \|\nabla\alpha\|_{L^2(\Omega)}^2 &= \int_\Omega \nabla\alpha \cdot (p - p_h)\, dx \\ &= \int_\Omega \alpha \operatorname{div}(p_h - p)\, dx. \end{aligned} \tag{22}$$

Note that there are no boundary terms because α vanishes on $\partial\Omega$. Since $\operatorname{div}(p - p_h) = g_{\mathcal{T}} - g$ and with the piecewise constant integral mean $\alpha_{\mathcal{T}}$ of α there holds

$$\begin{aligned}\int_\Omega \alpha \operatorname{div}(p_h - p)\,dx &= \int_\Omega (g - g_{\mathcal{T}})(\alpha - \alpha_{\mathcal{T}})\,dx \\ &\le \operatorname{osc}(g, \mathcal{T}) \left\| h_{\mathcal{T}}^{-1}(\alpha - \alpha_{\mathcal{T}}) \right\|_{L^2(\Omega)}.\end{aligned} \tag{23}$$

A Poincaré inequality proves

$$\left\| h_{\mathcal{T}}^{-1}(\alpha - \alpha_{\mathcal{T}}) \right\|_{L^2(\Omega)} \lesssim \|\nabla\alpha\|_{L^2(\Omega)}. \tag{24}$$

The combination of (22)-(24) proves (21).

Define β_h as some interpolation of β such that $\beta_h \in P_1(\mathcal{T})$ and $\operatorname{Curl}\beta_h \in P_0(\mathcal{T})^2$ and $\operatorname{Curl}\beta_h \perp \nabla H_0^1(\Omega)$, i.e.,

$$\int_\Omega \nabla v \cdot \operatorname{Curl}\beta_h\,dx = 0 \quad \text{for all } v \in H_0^1(\Omega),$$

whence $\operatorname{div}\operatorname{Curl}\beta_h = 0$.

The second key observation reads

$$\int_\Omega \operatorname{Curl}\beta \cdot \operatorname{Curl}\beta_h\,dx = 0. \tag{25}$$

By orthogonality of $\nabla\alpha$ and $\operatorname{Curl}\beta_h$ there holds

$$\int_\Omega \operatorname{Curl}\beta \cdot \operatorname{Curl}\beta_h\,dx = -\int_\Omega (p - p_h) \cdot \operatorname{Curl}\beta_h\,dx.$$

An integration by parts for $p = \nabla u$ and the discrete equations (11) for $q_h = \operatorname{Curl}\beta_h \in RT_0(\mathcal{T})$ with $\operatorname{div} q_h = 0$ show that the last expression equals

$$-\int_\Omega (p - p_h) \cdot \operatorname{Curl}\beta_h\,dx = \int_\Omega (u - u_h)\operatorname{div}\operatorname{Curl}\beta_h\,dx = 0.$$

Altogether, one deduces (25).

The third key assertion reads

$$\|\operatorname{Curl}\beta\|_{L^2(\Omega)} \lesssim \eta. \tag{26}$$

From (25) and the orthogonality of $\nabla\alpha$ and $\operatorname{Curl}\beta$, one has

$$\|\operatorname{Curl}\beta\|_{L^2(\Omega)}^2 = \int_\Omega (p - p_h) \cdot \operatorname{Curl}(\beta - \beta_h)\,dx.$$

This holds for all $\beta_h \in \mathcal{S}^1(\mathcal{T})$ which we may choose as some Clément-type approximation of β which can be proven Clément (1975); Carstensen and Funken (2000) to satisfy

$$\|\beta-\beta_h\|_{L^2(T)} \lesssim h_T \|\nabla\beta\|_{L^2(\omega_T)} \quad \text{and} \quad \|\beta-\beta_h\|_{L^2(E)} \lesssim h_E^{1/2} \|\nabla\beta\|_{L^2(\omega_E)}$$

for any $T \in \mathcal{T}$ and $E \in \mathcal{E}$ with local neighborhoods ω_T and ω_E. An elementwise integration by parts and the aforementioned approximation results of the Clément interpolation β_h show

$$\begin{aligned}
&\int_\Omega (p-p_h)\cdot \operatorname{Curl}(\beta-\beta_h)\,dx \\
&\quad = \sum_{T\in\mathcal{T}} \left(-\int_T \operatorname{curl}(p-p_h)(\beta-\beta_h)\,dx + \int_{\partial T}(p-p_h)\cdot\tau(\beta-\beta_h)\,ds\right) \\
&\quad \le \left(\sum_{T\in\mathcal{T}} h_T^2 \|\operatorname{curl} p_h\|_{L^2(T)}^2\right)^{1/2} \left(\sum_{T\in\mathcal{T}} h_T^{-2}\|\beta-\beta_h\|_{L^2(T)}^2\right)^{1/2} \\
&\qquad + \left(\sum_{E\in\mathcal{E}} h_E \int_E |[p_h]_E|^2\,ds\right)^{1/2} \left(\sum_{E\in\mathcal{E}} h_E^{-1}\|\beta-\beta_h\|_{L^2(E)}^2\right)^{1/2} \\
&\quad \lesssim \left(\sum_{T\in\mathcal{T}} h_T^2 \|\operatorname{curl} p_h\|_{L^2(T)}^2 + \sum_{E\in\mathcal{E}} h_E \int_E |[p_h]_E|^2\,ds\right)^{1/2} \|\nabla\beta\|_{L^2(\Omega)}.
\end{aligned}$$

Here τ is the tangential unit vector along the boundary ∂T. In RT_0 there holds $\operatorname{curl} p_h = 0$. Finally $\|\nabla\beta\|_{L^2(\Omega)} = \|\operatorname{Curl}\beta\|_{L^2(\Omega)}$ implies the assertion (26) in 2D.

The final key assertion reads

$$h_E^{1/2} \|[p_h]_E\|_{L^2(E)} \lesssim \|\operatorname{Curl}\beta\|_{L^2(\omega_E)}. \tag{27}$$

The proof is based on local test functions and inverse estimates. Details can be found in Carstensen (1997); Alonso (1996) and are omitted here.

Finally Theorem 5.2 follows from the combination of (21),(25)–(27). □

6 Unified Theory of A Posteriori Error Control

This lecture outlines a general theory for the a posteriori error analysis of mixed FEM to illustrate that the a posteriori estimators are in fact quite similar or even equivalent to conforming estimators.

Let $A : X \to X^*$ be a linear and bounded operator between the (real) Hilbert spaces X and its dual X^*. Suppose that A is surjective and injective such that A^{-1} is bounded as well. In particular, given $y \in X^*$, there exists a unique $x \in X$ such that

$$Ax = y. \tag{28}$$

Suppose we are given some finite element approximation $x_h \in X$ (possibly with some additional properties generated by the computational scheme that provided x_h). Then we address the issue of approximating the error $\|x - x_h\|$. Notice that the linear functional

$$\mathcal{R}es := y - Ax_h = A(x - x_h) \in X^* \tag{29}$$

is known or, at least, computable. Since the operator norms $\|A\| \approx 1 \approx \|A^{-1}\|$ of A and A^{-1} are uniformly bounded, there holds

$$\|x - x_h\|_X \approx \|A(x - x_h)\|_{X^*} \approx \|\mathcal{R}es\|_{X^*}. \tag{30}$$

Hence, any residual-based a posteriori error control means the approximation of lower and upper bounds of the dual norm $\|\mathcal{R}es\|_{X^*}$ of $\mathcal{R}es$. The primal variable $u \in H$ (e.g., the displacement field) is accompanied by a dual variable $p \in L$ (e.g., the flux or stress). The pair $(p, u) =: x \in L \times H =: X$ plays the role of the variable x above; L and H are Hilbert spaces. The linear bijective operator $A : X \to X^*$ is defined via a mixed framework, namely,

$$A(p, u)(q, v) := a(p, q) + b(p, v) + b(q, u) \tag{31}$$

for bounded bilinear forms $a : L \times L \to \mathbb{R}$ and $b : L \times H \to \mathbb{R}$. Given the right-hand sides $f \in L^*$ and $g \in H^*$ with $y \in X^*$ defined by $y(q, v) = f(q) + g(v)$, there exists a unique $x = (p, u) \in L \times H$ with (28). Let $x_h = (p_h, \tilde{u}_h) \in L \times H$ be some approximation to x and define

$$\mathcal{R}es := \mathcal{R}es_L + \mathcal{R}es_H$$

by (29), where the summands

$$\begin{aligned} \mathcal{R}es_L(q) &:= f(q) - a(p_h, q) - b(q, \tilde{u}_h) \quad \text{for } q \in L \quad \text{and} \\ \mathcal{R}es_H(v) &:= g(v) - b(p_h, v) \quad \text{for } v \in H \end{aligned} \tag{32}$$

act on their respective domains. The notation $\tilde{u}_h \in H$ here and below asserts that $\tilde{u}_h$ is a continuous and not necessarily a discrete function; the subindex h in $\tilde{u}_h$ refers to the fact that $\tilde{u}_h$ is closely related to some discontinuous $u_h \notin H$, computed by some MFEM, and is on our disposal. With (31)-(32), equivalence (30) reads

$$\|p - p_h\|_L + \|u - \tilde{u}_h\|_H \approx \|\mathcal{R}es_L\|_{L^*} + \|\mathcal{R}es_H\|_{H^*}. \tag{33}$$

This is the starting point of the unifying theory. The fact that $\tilde{u}_h$ has to belong to H, while u_h does not, is some crucial point in the sequel.

It should be notified clearly that (31) is a *primal mixed* formulation, also called hybrid in Brezzi and Fortin (1991), where L is *not* a subspace of $H(\mathrm{div};\Omega)$. This is because the derivatives act on $u \in H = H^1_0(\Omega)$. For the Poisson equation, the bilinear form b reads

$$b(q,v) = -\int_\Omega q \cdot \nabla v \, dx$$

(without an integration by parts as for the dual mixed formulation) and similar expressions hold for Stokes and Navier-Lamè equations described in Example 4.4 – 4.6.

This lecture emphasises that there is essentially only one type of residual that arise in a posteriori error control: Given some $g \in L^2(\Omega)$ and some $g_\mathcal{E} \in L^2(\cup\mathcal{E})$ one encounters the linear functional $\mathcal{R}es : H^1_0(\Omega) \to \mathbb{R}$, written $\mathcal{R}es \in H^{-1}(\Omega) \equiv H^1_0(\Omega)^*$, defined by

$$\mathcal{R}es(v) := \int_\Omega gv \, dx + \int_{\cup\mathcal{E}} g_\mathcal{E} v \, ds \tag{34}$$

with the Galerkin property

$$\mathcal{S}^1_0(\mathcal{T}) := P_1(\mathcal{T}) \cap H^1_0(\Omega) \subseteq \ker \mathcal{R}es.$$

The norm of $\mathcal{R}es$ enters as an explicit bound in the form of

$$\|\mathcal{R}es\| := \sup_{v \in H^1_0(\Omega)\setminus\{0\}} \mathcal{R}es(v)/\|\nabla v\|_{L^2(\Omega)}.$$

To fix ideas, suppose notation from Section 5. For any $p_h \in L := L^2(\Omega, \mathbb{R}^n)$ and $\tilde{u}_h \in H := H^1_0(\Omega)$ there holds (33) for $\mathcal{R}es_L \in L^*$ and $\mathcal{R}es_H \in H^*$ defined for all $q \in L$ and $v \in H$ by

$$\mathcal{R}es_L(q) := \int_\Omega q \cdot (\nabla \tilde{u}_h - p_h) \, dx \quad \text{and}$$

$$\mathcal{R}es_H(v) := -\int_\Omega gv \, dx + \int_\Omega p_h \cdot \nabla v \, dx.$$

Since the discrete solution u_h of the MFEM is, in general, not in H, the formulation above is applied to estimate $\|p - p_h\|_L + \|u - \tilde{u}_h\|_H$ for some $\tilde{u}_h \in H$ different from u_h. The evaluation of $\mathcal{R}es_H$ follows with a Poincaré inequality.

Let $g_{\mathcal{T}} \in P_0(\mathcal{T})$ denote the piecewise integral mean of $g \in H^1(\Omega)$. Then

$$g_{\mathcal{T}} - g = \operatorname{div}(p - p_h).$$

This and an integration by parts followed by the Poincaré inequality show

$$\begin{aligned}\mathcal{R}es_H(v) &= \int_\Omega p_h \cdot \nabla v\, dx - \int_\Omega gv\, dx \\ &= -\int_\Omega (g + \operatorname{div} p_h) v\, dx \\ &= -\int_\Omega (g - g_{\mathcal{T}})(v - v_{\mathcal{T}})\, dx \\ &\le \left\| h_{\mathcal{T}}^2 \nabla g \right\|_{L^2(\Omega)} \|\nabla v\|_{L^2(\Omega)}.\end{aligned}$$

Hence (h.o.t. abbreviates higher order terms)

$$\|\mathcal{R}es_H\|_{H^*} \le \left\| h_{\mathcal{T}}^2 \nabla g \right\|_{L^2(\Omega)} = \text{h.o.t.}$$

is of quadratic, whence of higher, order when compared to $\|p - p_h\|$ for RT_0. For the evaluation of $\|\mathcal{R}es_L\|_{L^*}$, recall that $\tilde{u}_h$ denotes an arbitrary (not necessarily discrete) element in H.

Definition 6.1. Given $p_h \in L^2(\Omega)^2$, define the linear functional

$$\mathcal{R}es_{NC} : H^1(\Omega) \to \mathbb{R} \quad \text{by} \quad \mathcal{R}es_{NC}(v) := \int_\Omega p_h \cdot \operatorname{Curl} v\, dx$$

for $v \in H^1(\Omega)$. Set

$$\|\mathcal{R}es_{NC}\| := \sup_{\substack{v \in H^1(\Omega), \\ \operatorname{Curl} v \neq 0}} \int_\Omega p_h \cdot \operatorname{Curl} v\, dx / \|\operatorname{Curl} v\|_L,$$

where NC refers to NonConforming (see Subsection 8.1 for more details).

Theorem 6.2. *For any p_h generated by an MFEM of Table 1 there holds*

$$\min_{\tilde{u}_h \in H} \|p_h - \nabla \tilde{u}_h\|_L = \|\mathcal{R}es_{NC}\| \quad \textit{and} \quad \mathcal{S}^1(\mathcal{T}) \subset \ker \mathcal{R}es_{NC}.$$

Proof. Given $v_h \in \mathcal{S}_1(\mathcal{T})$ one shows $\operatorname{Curl} v_h \in H(\operatorname{div};\Omega)$. Hence,

$$q_h := \operatorname{Curl} v_h \in P_0(\mathcal{T})^2 \cap H(\operatorname{div}, \Omega) \subset L_h.$$

The last inclusion holds for the MFEM spaces of Table 1. Thus, q_h may be considered in discrete MFEM and shows

$$\mathcal{R}es_{NC}(v_h) = \int_\Omega p_h \cdot \operatorname{Curl} v_h \, dx = -\int_\Omega u_h \operatorname{div} \operatorname{Curl} v_h \, dx = 0. \qquad \square$$

The main observation is that, since $p_h \cdot \operatorname{Curl} v = \tilde{p}_h \cdot \nabla v$ for some rotated $\tilde{p}_h$ and with integration by parts, $\mathcal{R}es_{NC}$ has the form (34). Thus the analysis of, e.g., Ainsworth and Oden (2000); Babuška and Strouboulis (2001); Verfürth (1989) applies and many alternative estimates are applicable. From the longer list of possible error estimators, only averaging and residual-type estimators (η_A and η_R) are mentioned.

$$\eta_A := \|h_\mathcal{T}^2 \nabla g\|_{L^2(\Omega)} + \Big\| p_h - \sum_{z \in \mathcal{N}} (A_z p_h)\varphi_z \Big\|_{L^2(\Omega)},$$
$$\eta_R^{NC} := \Big(\sum_{E \in \mathcal{E}} h_E \, \big\| [p_h]_E \cdot \tau_E \big\|_{L^2(E)}^2 \Big)^{1/2},$$

where $A_z : L^2(\Omega)^2 \to \mathbb{R}^2$ is an averaging operator at node $z \in \mathcal{N}$, e.g.,

$$A_z \, p_h := \frac{1}{|\omega_z|} \int_{\omega_z} p_h \, dx,$$

for $\omega_z := \{x \in \Omega \, : \, 0 < \phi_z(x)\}$. Then based on Theorem 6.2,

$$\|\mathcal{R}es_{NC}\| \approx \eta_R^{NC} \approx \eta_A.$$

In the same unified spirit, other applications can be viewed. For the Stokes problem, the paper Carstensen (2005) analyses general estimates for the stress

$$\sigma := 2\mu \, \varepsilon(u) - p\mathbf{1} \in L := L^2(\Omega; \mathbb{R}^{2\times 2}_{\mathrm{sym}})$$

and for $u_h \in H^1(\mathcal{T})^2$ and $p_h \in L^2_0(\Omega)$

$$\sigma_h := 2\mu \, \varepsilon_\mathcal{T}(u_h) - p_h\mathbf{1} \in L.$$

For linear elasticity, Carstensen (2005) analyses λ-robust approximations with a mixed finite element method and establishes, for any $\tilde{u}_h \in H^1_0(\Omega)^2$, that

$$\|\sigma - \operatorname{sym} \sigma_h\|_L + \|u - \tilde{u}_h\|_H \approx \|\varepsilon(\tilde{u}_h) - \mathbb{C}^{-1} \operatorname{sym}(\sigma_h)\|_L + \|\mathcal{R}es_H\|_{H^*}.$$

This estimate leads to

$$\|\sigma - \sigma_h\|_L \approx \|\mathcal{R}es_{NC}\|_{H^*} + \|\mathcal{R}es_H\|_{H^*} + \|\operatorname{skew}(\sigma_h)\|_L$$

and hence to similar estimate as in the literature.

7 Error Reduction in AMFEM

This lecture is devoted to the error reduction property of adaptive mixed finite element methods (AMFEM) for the Poisson Problem. Unlike for conforming AFEM, there is no further restriction (like inner-node properties etc.) in REFINE. The results of this lecture are taken from Carstensen and Hoppe (2006).

Throughout this lecture $\mathcal{T}_H$ and $\mathcal{T}_h$ denote two shape-regular triangulations of Ω into triangles where $\mathcal{T}_h$ is some refinement of $\mathcal{T}_H$ such that the refinement $T|_{\mathcal{T}_h} := \{K \in \mathcal{T}_h : K \subseteq T\}$ of each element T in $\mathcal{T}_H$ is depicted in Figure 8. Moreover, let $p_H \in RT_0(\mathcal{T}_H)$ denote the discrete MFEM solution on the coarse triangulation $\mathcal{T}_H$. The subindices H and h refer to the coarse and fine triangulation $\mathcal{T}_H$ and $\mathcal{T}_h$, respectively, e.g., p, p_h, p_H denote the exact and discrete fluxes of RT_0 AFEM of Subsection 3.5 with respect to two consecutive meshes $\mathcal{T}_h$ and $\mathcal{T}_H$.

Theorem 7.1 (Error Reduction Property). *There holds*

$$\|p - p_h\|^2_{L^2(\Omega)} \le \rho \|p - p_H\|^2_{L^2(\Omega)} + C \operatorname{osc}_H \|H f_H\|_{L^2(\Omega)}$$

for some positive constants $\rho < 1$ and C which exclusively depend on the minimal angle in $\mathcal{T}_H$ and $\mathcal{T}_h$ and with H, $f_H \in P_0(\mathcal{T}_H)$ defined by $f_H := f_{\mathcal{T}_H}$ and $H|_T := \operatorname{diam}(T)$ for $T \in \mathcal{T}_H$, and

$$\operatorname{osc}_H := \Big(\sum_{E \in \mathcal{E}_H} h_E^2 \|f - f_{\omega_E}\|^2_{L^2(\omega_E)} \Big)^{1/2}.$$

The remaining part of this lecture is devoted to the proof of Theorem 7.1 and based on some orthogonality and the discrete local efficiency. Throughout, the following notation for any interior edge is employed. Suppose that $E = \partial T_+ \cap \partial T_- \in \mathcal{E}_H$ is an edge in $\mathcal{T}_H$ (shared by the triangles $T_+, T_- \in \mathcal{T}_H$ from Figure 5) and bisected in the refinement, i.e., $E = E_1 \cup E_2 \notin \mathcal{E}_h$ and $\{\operatorname{mid}(E)\} = E_1 \cap E_2 \in \mathcal{N}_h$ for two distinct $E_1, E_2 \in \mathcal{E}_h$. Observe that $[p_H] \cdot \nu_E = 0$ for the unit normal vector $\nu_E \perp E$ since $p_H \in H(\operatorname{div}, \Omega)$.

Therefore, the jump $[p_H]$ along $E = T_+ \cap T_-$ in the tangential direction $\tau_E \perp \nu_E$ (abbreviated $[p_H]_E$ in the sequel) satisfies that

$$\|[p_H]\|_{L^2(E)} = \|[p_H] \cdot \tau_E\|_{L^2(E)}.$$

Since $[p_H] \cdot \tau_E$ is a scalar affine function along the edge E, it reads

$$([p_H] \cdot \tau_E)(x) = \alpha + \beta \cdot (x - \mathrm{mid}(E)) \quad \text{for all } x \in E \subseteq \mathbb{R}^2$$

with fixed $\alpha = \int_E [p_H] \cdot \tau_E \, ds / h_E \in \mathbb{R}$ and $\beta \in \mathbb{R}^2$, $h_E = \mathrm{diam}(E)$.

Proposition 7.2 (Discrete Local Efficiency). *There holds*

$$h_E^{1/2} \|[p_h]\|_{L^2(E)} \lesssim \|p_h - p_H\|_{L^2(\omega_E)} + h_E \|f - f_{\omega_E}\|_{L^2(\omega_E)}.$$

Proof. The first assertion reads

$$h_E \, |\alpha| \lesssim \|p_h - p_H\|_{L^2(\omega_E)}. \tag{35}$$

To proof (35), let φ_E denote the nodal basis function in the conforming P_1 FEM space with respect to the node mid(E) and in the fine mesh $\mathcal{T}_h$. Since φ_E is continuous along E, $q_h := \mathrm{Curl}\, \varphi_E$ belongs to $P_0(\mathcal{T}_h)^2 \cap H(\mathrm{div}, \Omega) \subset RT_0(\mathcal{T}_h)$ with $\mathrm{div}\, q_h \equiv 0$. Since $q_h \in RT_0(\mathcal{T}_h)$ is an admissible test function, the discrete MFEM problem with respect to the fine mesh $\mathcal{T}_h$ leads to

$$(p_h, q_h)_{L^2(\Omega)} = 0.$$

Since $\int_E \varphi_E \, ds = h_E/2$ and with an elementwise integration by parts in the last step, we have

$$\int_E \alpha \varphi_E \, ds = \int_E [p_H] \cdot \tau_E \varphi_E \, ds = (p_H, q_h)_{L^2(\Omega)}.$$

Altogether, one obtains the key identity

$$\alpha \int_E \varphi_E \, ds = (p_H - p_h, q_h)_{L^2(\Omega)}.$$

The shape regularity allows the estimate $\|q_h\|_{L^2(\omega_E)} \lesssim 1$. Hence, the foregoing key identity leads to the assertion (35). In fact,

$$\begin{aligned} h_E^2 \, |\alpha|^2 = h_E \|\alpha\|_{L^2(E)}^2 &= h_E^2 \Big(\int_E \varphi_E \, ds \Big)^{-2} \Big(\alpha \int_E \varphi_E \, ds \Big)^2 \\ &\lesssim (p_H - p_h, q_h)_{L^2(\Omega)}^2 \le \|q_h\|_{L^2(\omega_E)}^2 \|p_h - p_H\|_{L^2(\omega_E)}^2 \\ &\lesssim \|p_h - p_H\|_{L^2(\omega_E)}^2. \end{aligned}$$

The second assertion reads

$$|\beta| \le \frac{1}{2}(|T_+|^{-1} + |T_-|^{-1})^{1/2}\|f - f_{\omega_E}\|_{L^2(\omega_E)}. \tag{36}$$

Indeed, the differences of the representation formula (42) of Theorem 8.1 below for $x \in E$ lead to

$$\beta = \frac{1}{2}(f_{T_-} - f_{T_+})\,\tau_E \in \mathbb{R}^2.$$

The piecewise constant function

$$g(x) := \begin{cases} -|T_+|^{-1} & \text{for } x \in T_+, \\ +|T_-|^{-1} & \text{for } x \in T_-, \\ 0 & \text{for } x \notin \overline{\omega_E} \end{cases}$$

satisfies $\int_{\omega_E} g(x)\,dx = 0$. The definition of the piecewise integral mean $f_{T_\pm} := \fint_{T_\pm} f(x)\,dx$ then implies the identity

$$f_{T_-} - f_{T_+} = (g, f)_{L^2(\Omega)}.$$

Since $(g,1)_{L^2(\Omega)} = 0$ and $f_{\omega_E} := |\omega_E|^{-1}\int_{\omega_E} f(x)\,dx$ is constant on ω_E,

$$f_{T_-} - f_{T_+} = (g, f - f_{\omega_E})_{L^2(\Omega)}.$$

A Cauchy's inequality and $\|g\|^2_{L^2(\omega_E)} = |T_+|^{-1} + |T_-|^{-1}$ lead to (36).

Since α and $\beta\cdot(\cdot - \mathrm{mid}(E))$ (which stands for $\beta\cdot(x - \mathrm{mid}(E))$ as a function of x) are $L^2(E)$ orthogonal, there holds

$$\begin{aligned} h_E\|[p_H]\|^2_{L^2(E)} &= h_E\|[p_H]\cdot\tau_E\|^2_{L^2(E)} \\ &= h_E\|\alpha\|^2_{L^2(E)} + h_E\|\beta\cdot(\cdot - \mathrm{mid}(E))\|^2_{L^2(E)} \\ &\lesssim \|p_h - p_H\|^2_{L^2(\omega_E)} \\ &\qquad + h_E^{-1}\|\cdot - \mathrm{mid}(E)\|^2_{L^2(E)}\|f - f_{\omega_E}\|^2_{L^2(\omega_E)}. \end{aligned}$$

This proves Proposition 7.2. □

Throughout the remaining part of this lecture, $H \in P_0(\mathcal{T}_H)$ denotes the mesh-size, $H|_T = h_T = \mathrm{diam}(T)$ for any $T \in \mathcal{T}_H$. Notice that $\|p - p_h\|_{L^2(\Omega)} + \|p - p_H\|_{L^2(\Omega)} + \|Hf_H\|_{L^2(\Omega)}$ are first-order terms which compare with the flux error in the MFEM, while

$$\|H(f_h - f_H)\| \le \|H(f - f_H)\| + \|h(f - f_h)\| \le \mathrm{osc}(f)$$

is of higher order if the right-hand side f is smooth.

Proposition 7.3 (Quasi-Orthogonality). *There holds*

$$|(p-p_h,p_H-p_h)_{L^2(\Omega)}| \lesssim \Big(\|p-p_h\|_{L^2(\Omega)} + \|p-p_H\|_{L^2(\Omega)} + \|Hf_H\|_{L^2(\Omega)}\Big) \|H(f_h-f_H)\|_{L^2(\Omega)}. \tag{37}$$

Proof. For $p_h^{NC} = \nabla_h u_h^{NC}$ [from the nonconforming Crouzeix-Raviart finite element scheme in Subsection 8.1] the first key assertion reads

$$|(p-p_h^{NC},p_H-p_h)_{L^2(\Omega)}| \lesssim \Big(\|p-p_h\|_{L^2(\Omega)} + \|p-p_H\|_{L^2(\Omega)} + \|hf_h\|_{L^2(\Omega)} + \|Hf_H\|_{L^2(\Omega)}\Big) \|H(f_h-f_H)\|_{L^2(\Omega)}. \tag{38}$$

In fact, since $p=\nabla u$, $-\operatorname{div} p_H = f_H$, and $-\operatorname{div} p_h = f_h$, an elementwise integration by parts shows

$$(p-p_h^{NC},p_H-p_h)_{L^2(\Omega)} = (u-u_h^{NC},f_H-f_h)_{L^2(\Omega)}.$$

Therein, the edge contributions vanish: Given any $E\in\mathcal{E}_h$ the resulting boundary term over E reads

$$\int_E [u-u_h^{NC}]\,(p_H-p_h)\cdot\nu_E\,ds$$

and equals zero because of $\int_E [u-u_h^{NC}]\,ds = 0$ by construction of $\mathcal{S}_0^{1,NC}$ and since $p_H\cdot\nu_E$ and $p_h\cdot\nu_E$ are continuous from both sides of E and constant along E.

To estimate $(u-u_h^{NC},f_H-f_h)_{L^2(\Omega)}$ notice that $\int_T(f_H-f_h)\,dx = 0$ for any $T\in\mathcal{T}_H$. Hence, for some $\overline{e}_H^{NC}\in P_0(\mathcal{T}_H)$ with $\overline{e}_H^{NC}|_T := \fint_T(u(x)-u_H^{NC}(x))\,dx$ and $e_H^{NC} := u-u_H^{NC}$, a Poincaré inequality on T shows in total

$$\begin{aligned}|(u-u_H^{NC},f_H-f_h)_{L^2(\Omega)}| &= |(e_H^{NC}-\overline{e}_H^{NC},f_H-f_h)_{L^2(\Omega)}|\\ &\le \|p-p_H^{NC}\|_{L^2(\Omega)}\|H(f_h-f_H)\|_{L^2(\Omega)}.\end{aligned}$$

The remaining term reads $(u_H^{NC}-u_h^{NC},f_H-f_h)_{L^2(\Omega)}$ and is analysed separately for each $T\in\mathcal{T}_H$. In fact, let

$$\mathcal{S}_0^{1,NC}(T) := \{v_h|_T : v_h\in P_1(\mathcal{T}_h|_T) \text{ continuous at } \operatorname{mid}(E) \text{ for all } E\in\mathcal{E}_h\}$$

and notice $u_H^{NC}-u_h^{NC}\in\mathcal{S}_0^{1,NC}(T)$. Moreover, for any $v_h\in\mathcal{S}_0^{1,NC}(T)$ set

$$\varrho_1(v_h) := \min_{w\in\mathbb{R}}\|v_h-w\|_{L^2(T)} \quad\text{and}\quad \varrho_2(v_h) := h_T\|\nabla_h v_h\|_{L^2(T)}.$$

This defines two norms ϱ_1, ϱ_2 on the finite-dimensional space $\mathcal{S}_0^{1,NC}(\mathcal{T})/\mathbb{R}$. Consequently, $\varrho_1 \approx \varrho_2$. Therein, the equivalence constants are independent of h_T according to a scaling argument (transform to a reference triangle T_{ref} first and notice that there exists a finite number of possible refinements only, compute the constants, and transform back). In particular, for some average $c := \int_T (u_H^{NC} - u_h^{NC})\, dx/|T|$,

$$\begin{aligned}|(u_h^{NC} - u_H^{NC}, f_h - f_H)_{L^2(T)}| &= |(u_H^{NC} - u_h^{NC} - c, f_H - f_h)_{L^2(\Gamma)}| \\ &\leq \varrho_1(u_H^{NC} - u_h^{NC}) \|f_h - f_H\|_{L^2(T)} \\ &\lesssim \|\nabla_h(u_H^{NC} - u_h^{NC})\|_{L^2(T)} \\ &\qquad \|h_T(f_h - f_H)\|_{L^2(T)}.\end{aligned}$$

The sum over all $T \in \mathcal{T}_H$ shows that

$$|(u_h^{NC} - u_H^{NC}, f_h - f_H)_{L^2(\Omega)}| \lesssim \|p_h^{NC} - p_H^{NC}\|_{L^2(\Omega)} \, \|H(f_h - f_H)\|_{L^2(\Omega)}.$$

An immediate consequence of the representation formula (42) from below is that

$$\Big| \|p - p_H^{NC}\|_{L^2(\Omega)} - \|p - p_H\|_{L^2(\Omega)} \Big|^2 \leq \sum_{T \in \mathcal{T}_H} |f_T|^2 \|\cdot - x_T\|^2_{L^2(T)} \leq \|H f_H\|^2_{L^2(\Omega)}.$$

There also holds an analog estimate with H replaced by h. The combination of the two aforementioned estimates shows (38).

The second key assumption reads

$$|(p_h^{NC} - p_h, p_H - p_h)_{L^2(\Omega)}| \lesssim \|h f_h\|_{L^2(\Omega)} \, \|h(f_h - f_H)\|_{L^2(\Omega)}. \tag{39}$$

To see this, let $x_H \in P_0(\mathcal{T}_H; \mathbb{R}^2)$ and $x_h \in P_0(\mathcal{T}_h; \mathbb{R}^2)$ denote the piecewise center of inertia, e.g., $x_H|_T := \text{mid}(T)$ for $T \in \mathcal{T}_H$. Then, the representation formula (42) reads

$$p_H(x) - p_H^{NC}(x) = -\frac{1}{2} f_H\,(x - x_H) \quad \text{for } x \in \Omega$$

plus a corresponding equation with H replaced by h. Then,

$$\begin{aligned}(p_h^{NC} - p_h, p_H - p_h)_{L^2(\Omega)} &= \frac{1}{2}\,(f_h(\cdot - x_h), p_H^{NC} - p_h^{NC})_{L^2(\Omega)} \\ &\quad + \frac{1}{4}(f_h(\cdot - x_h), f_h(\cdot - x_h) - f_H(\cdot - x_H))_{L^2(\Omega)}.\end{aligned}$$

The first term on the right-hand side vanishes because $(p_H^{NC} - p_h^{NC})|_T$ is constant and $\int_T (x - x_T)\, dx = 0$ for each $T \in \mathcal{T}_h$. The same argument shows $(f_h(\cdot - x_h), x_H - x_h)_{L^2(\Omega)} = 0$. There remains

$$4(p_h^{NC} - p_h, p_H - p_h)_{L^2(\Omega)} = (f_h(\cdot - x_h), (f_h - f_H)(\cdot - x_h))_{L^2(\Omega)}.$$

An elementwise Cauchy inequality in the previous identity leads to (39).

Finally, the estimates (38) and (39) yield Proposition 7.3. □

Proof of Theorem 7.1. One key assumption is the use of the bulk criterion in MARK to ensure

$$\eta^2 := \sum_{E \in \mathcal{E}_H} h_E \|[p_H]\|_{L^2(E)}^2 \lesssim \sum_{E \in \mathcal{M}} h_E \|[p_H]\|_{L^2(E)}^2 \tag{40}$$

for the set $\mathcal{M}$ of marked edges. Moreover, each edge $E \in \mathcal{M}$ is bisected in the refinement of $\mathcal{T}_H$ to design $\mathcal{T}_h$. The reliability from Theorem 5.2 and the bulk criterion (40) lead to

$$\|p - p_H\|_{L^2(\Omega)}^2 \lesssim \eta^2 + \mathrm{osc}_H^2 \lesssim \sum_{E \in \mathcal{M}} h_E \|[p_H]\|_{L^2(E)}^2 + \mathrm{osc}_H^2 .$$

The discrete local efficiency of Theorem 7.2 plus the finite overlap of the edge-patches $(\omega_E : E \in \mathcal{E}_H)$ show

$$\|p - p_H\|_{L^2(\Omega)}^2 \leq c_1 \|p_h - p_H\|_{L^2(\Omega)}^2 + c_1 \,\mathrm{osc}_H^2$$

with some constant c_1. On the other hand,

$$\|p_h - p_H\|_{L^2(\Omega)}^2 = \|p - p_H\|_{L^2(\Omega)}^2 - \|p - p_h\|_{L^2(\Omega)}^2 - 2(p - p_h, p_h - p_H)_{L^2(\Omega)}$$

and the last term can be bounded with Theorem 7.3. With some constant $c_2 > 0$, this leads to

$$\begin{aligned}\|p_h - p_H\|_{L^2(\Omega)}^2 \leq \|p - p_H\|_{L^2(\Omega)}^2 &- \|p - p_h\|_{L^2(\Omega)}^2 \\ &+ c_2(\|p - p_h\|_{L^2(\Omega)} + \|p - p_H\|_{L^2(\Omega)} \\ &\quad + \|H f_H\|_{L^2(\Omega)}) \,\mathrm{osc}_H .\end{aligned}$$

The combination with the preceding inequality plus a Young inequality yield

$$\begin{aligned}
c_1 \|p - p_h\|^2_{L^2(\Omega)} &\le (c_1 - 1)\,\|p - p_H\|^2_{L^2(\Omega)} + c_1 \operatorname{osc}^2_H \\
&\quad + c_1 c_2 \Big(\|p - p_h\|_{L^2(\Omega)} + \|p - p_H\|_{L^2(\Omega)} \\
&\qquad + \|H f_H\|_{L^2(\Omega)}\Big) \operatorname{osc}_H \\
&\le \frac{1}{4}\|p - p_h\|^2_{L^2(\Omega)} + \Big(c_1 - \frac{1}{2}\Big)\|p - p_H\|^2_{L^2(\Omega)} \\
&\quad + c_2 \Big(\|H f_h\|_{L^2(\Omega)} + \operatorname{osc}_H\Big) \operatorname{osc}_H .
\end{aligned}$$

This proves

$$\Big(c_1 - \frac{1}{4}\Big)\,\|p - p_h\|^2_{L^2(\Omega)} \le \Big(c_1 - \frac{1}{2}\Big)\,\|p - p_H\|^2_{L^2(\Omega)} + c_2\,\|H f_h\|_{L^2(\Omega)} \operatorname{osc}_H$$

and so the theorem with $\rho = (c_2 - \frac{1}{2})/(c_1 - \frac{1}{4})$. □

8 Miscellaneous Remarks

This lecture remarks on the relation of nonconforming finite element methods to mixed FEM, the linear elastic problem and the PEERS for robust approximation, states the Reissner-Mindlin problem and comments on goal oriented adaptive mesh-refinement.

8.1 ANCFEM

This subsection is devoted to the flux approximation of the Raviart-Thomas MFEM obtained from nonconforming Crouzeix-Raviart finite elements. The key identity reads

$$p_h^{NC}(x) = \nabla_{\mathcal{T}}\, u_{NC} - f_{\mathcal{T}}(x - x_T)/2 \quad \text{for } x \in T \in \mathcal{T}$$

with the discrete flux p_h^{NC} from the lowest-order RT MFEM, the discrete flux $\nabla_{\mathcal{T}}\, u_h^{NC}$ from the P_1 nonconforming FE approximation and the center of gravity $x_T = \operatorname{mid}(T)$ of T. Set

$$\begin{aligned}
\mathcal{S}^{1,NC}(\mathcal{T}) &= \{v_h \in P_1(\mathcal{T}) :\ v_h \text{ are continuous in all midpoints } z_E \\
&\qquad\qquad \text{of edges } E \in \mathcal{E}_\Omega\}, \\
\mathcal{S}_0^{1,NC}(\mathcal{T}) &= \{v_h \in \mathcal{S}^{1,NC}(\mathcal{T}) :\ v_h(z_E) = 0 \text{ for all } E \in \mathcal{E}_\Gamma\},
\end{aligned}$$

where $\mathcal{E}_\Gamma := \{E \in \mathcal{E} :\ E \subseteq \partial\Omega\}$ and $\mathcal{E}_\Omega = \mathcal{E}\setminus\mathcal{E}_\Gamma$. Since $\mathcal{S}_0^{1,NC} \not\subset H^1(\Omega)$, the distributional gradient of $v_h \in \mathcal{S}_0^{1,NC}$ is different from its elementwise gradient $\nabla_{\mathcal{T}} v_h \in P_0(\mathcal{T})^2$. The discrete problem reads: *Find* $u_h^{NC} \in \mathcal{S}_0^{1,NC}(\mathcal{T})$

such that

$$\int_\Omega \nabla_\mathcal{T}\, u_h^{NC} \cdot \nabla_\mathcal{T}\, v_h\, dx = \int_\Omega f_\mathcal{T}\, v_h\, dx \quad \textit{for all } v_h \in \mathcal{S}_0^{1,NC}(\mathcal{T}). \tag{41}$$

It is well known that (41) has a unique solution u_h^{NC}.

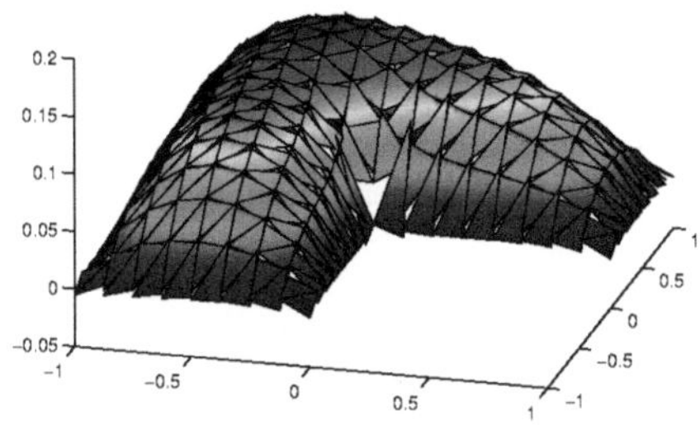

Figure 11. Visualisation of u_h^{NC} for the data and the uniform triangulation as in Figure 9.

Theorem 8.1 (Marini). *Let $p_h \in RT_0(\mathcal{T})$ solve the mixed system with the right-hand side $f \in L^2(\Omega)$. Let $u_h^{NC} \subset \mathcal{S}_0^{1,NC}(\mathcal{T})$ solve (41) with $f_\mathcal{T} = -\operatorname{div} p_h$. Set $b_h(x) = f_\mathcal{T}(x - \operatorname{mid}(T))/n$ for $x \in T \in \mathcal{T}$. Then, there holds*

$$p_h = \nabla_\mathcal{T}\, u_h^{NC} + b_h. \tag{42}$$

Proof. The proof dates back to Marini (1985). The displayed version is proven in Bahriawati and Carstensen (2005). □

The local stiffness matrix of the P_1 nonconforming finite element method essentially equals the conforming one. In fact, for any $T \in \mathcal{T}$ let M^{NC} and M denote the local stiffness matrix of P_1 nonconforming finite element and P_1 finite element, respectively. Then there holds

$$M^{NC} = 4M.$$

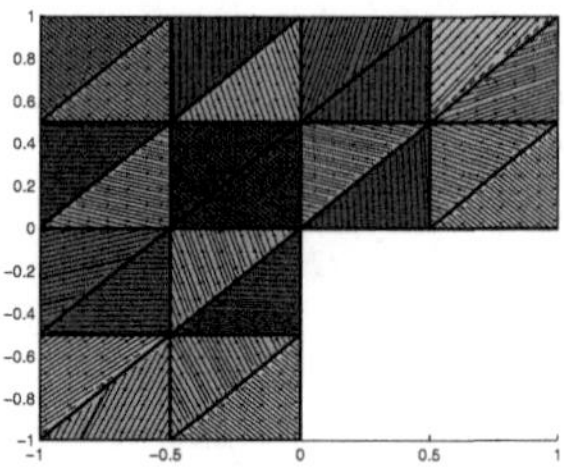

(a) $\nabla_{\mathcal{T}} u_h^{NC} \in P_0(\mathcal{T})$

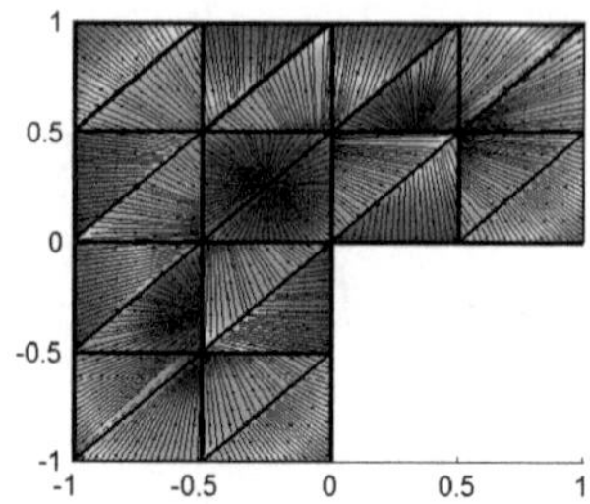

(b) $p_h \in RT_0(\mathcal{T})$

Figure 12. Illustration of Theorem 8.1

8.2 PEERS

The first MFEM for linear elasticity in history involved a non-symmetric stress tensor in $2D$ and hence was called **p**lane **e**lastic **e**lements with **r**educed **s**ymmetries or PEERS for brevity. The definition of the finite element spaces involves the bubble function $b_T = \lambda_1\lambda_2\lambda_3$ on a triangle $T \in \mathcal{T}$, where $\lambda_1, \lambda_2, \lambda_3$ are the barycentric coordinates of the triangle T in the regular triangulation $\mathcal{T}$. The PEERS is based on the following function spaces

$$\begin{aligned}
\mathcal{U} \times \mathcal{W} &= L^2(\Omega;\mathbb{R}^2) \times L^2(\Omega;\mathbb{R}^{2\times 2}_{\text{skew}}),\\
\mathcal{U}_h &= \{v_h \in \mathcal{U} : \forall T \in \mathcal{T}\; v_h|_T \in P_0(T;\mathbb{R}^2)\},\\
\mathcal{W}_h &= \{\gamma_h \in \mathcal{W} \cap C_0(\Omega;\mathbb{R}^{2\times 2}_{\text{skew}}) : \forall T \in \mathcal{T},\ \gamma_h|_T \in P_1(T;\mathbb{R}^{2\times 2}_{\text{skew}})\},\\
\Sigma_h &= \{\sigma_h \in L^2(\Omega;\mathbb{R}^{2\times 2}) : \operatorname{div}\sigma_h \in \mathcal{U},\ \forall T \in \mathcal{T},\\
&\qquad \sigma_h|_T \in RT_0(T) \oplus B_0(T)\}.
\end{aligned}$$

Here, RT_0 is the Raviart-Thomas space of lowest degree in two components, and

$$\begin{aligned}
RT_0(T) &= \{\sigma \in L^2(T;\mathbb{R}^{2\times 2}) : \sigma = \tau + a \otimes x,\ \tau \in \mathbb{R}^{2\times 2}, a \in \mathbb{R}^2\},\\
B_0(T) &= \{\sigma \in L^2(T;\mathbb{R}^{2\times 2}) : \sigma = a \otimes \operatorname{Curl} b_T,\ a \in \mathbb{R}^2\},\\
BDM_k(\Omega) &= \{\sigma_h \in L^2(\Omega;\mathbb{R}^{2\times 2}) : \operatorname{div}\sigma_h \in \mathcal{U}, \sigma_h|_T \in P_k(T;\mathbb{R}^{2\times 2})\}.
\end{aligned}$$

The higher-order methods $BDMS_k$ are based on the following function spaces (for $k \geq 2$) defined for $k \geq 2$ by

$$\begin{aligned}
\mathcal{U}_h &= \{v_h \in \mathcal{U} : \forall T \in \mathcal{T}\, v_h|_T \in P_{k-1}(T;\mathbb{R}^2)\},\\
\mathcal{W}_h &= \{\gamma_h \in \mathcal{W} : \forall T \in \mathcal{T}\, \gamma_h|_T \in P_k(T;\mathbb{R}^{2\times 2}_{\text{skew}})\},\\
\Sigma_h &= BDM_k(\Omega) \oplus \{\sigma_h \in L^2(\Omega;\mathbb{R}^{2\times 2}) : \operatorname{div}\sigma_h \in \mathcal{U},\\
&\qquad \forall T \in \mathcal{T}\, \sigma_h|_T \in P_k(T;\mathbb{R}^{2\times 2}) \oplus B_{k-1}(T)\},\\
B_{k-1}(T) &= \{\sigma \in L^2(T;\mathbb{R}^{2\times 2}) : \sigma = \operatorname{Curl}(b_T w), w \in P_{k-1}(T;\mathbb{R}^2)\}.
\end{aligned}$$

It is shown in Arnold et al. (1984); Stenberg (1988) that, for sufficiently small mesh-sizes there exists a unique discrete solution $(\sigma_h, u_h, \gamma_h) \in \Sigma_h \times \mathcal{U}_h \times \mathcal{W}_h$ of

$$a(\sigma_h, \tau_h) + b(\tau_h; u_h, \gamma_h) = 0 \quad \text{and} \quad b(\sigma_h; v_h, \eta_h) = g(v)$$

for all $(\tau_h, v_h, \eta_h) \in \Sigma_h \times \mathcal{U}_h \times \mathcal{W}_h$, and for $a(\sigma, \tau)$ and g from (7) with

$$b(\sigma; u, \gamma) := \int_\Omega (u \cdot \operatorname{div}\sigma + \sigma : \gamma)\, dx.$$

Moreover, with the exact solution u of (6) plus $\sigma := \mathbb{C}\varepsilon(u)$ and $\gamma := \operatorname{skew}(\nabla u)$, i.e., $\nabla u = \varepsilon(u) + \gamma$, there holds quasi-optimal convergence (with arbitrary $(\tau_h, v_h, \eta_h) \in \Sigma_h \times U_h \times W_h$ on the right hand side)

$$\begin{aligned}
&\|u - u_h\|_{L^2(\Omega)} + \|\sigma - \sigma_h\|_{H(\operatorname{div},\Omega)}\\
&\qquad \lesssim \min_{(\tau_h, v_h, \eta_h)} \Big(\|u - v_h\|_{L^2(\Omega)} + \|\sigma - \tau_h\|_{H(\operatorname{div},\Omega)} + \|\gamma - \eta_h\|_{L^2(\Omega)} \Big).
\end{aligned}$$

Proofs are given in Arnold et al. (1984); Stenberg (1988) and involve the LBB condition via some macro-element technique.

The a posteriori error estimates from Carstensen and Dolzmann (1998); Carstensen et al. (2000) established the robust reliable and efficient error estimator η with

$$\|u - u_h\|_{L^2(\Omega)} + \|\gamma - \gamma_h\|_{L^2(\Omega)} + \|\mathbb{C}^{-1/2}\sigma - \sigma_h\|_{L^2(\Omega)} \approx \eta + \operatorname{osc}(g, \mathcal{T}),$$

where, for each $T \in \mathcal{T}$ and $E \in \mathcal{E}$,

$$\begin{aligned}
\eta_T^2 &= h_T^2\, \|\operatorname{div}\varepsilon\|^2_{L^2(T)} + h_T^2\, \|\operatorname{curl}(\mathbb{C}^{-1}\sigma_h + \gamma_h)\|^2_{L^2(T)}\\
&\quad + h_T^2 \inf_{v_h \in \mathcal{U}_h} \|\mathbb{C}^{-1}\sigma_h + \gamma_h - \nabla v_h\|^2_{L^2(T)} + \|\operatorname{skew}(\sigma_h)\|^2_{L^2(T)},\\
\eta_E^2 &= h_E\, \|J((\mathbb{C}^{-1}\sigma_h + \gamma_h)t)\|^2_{L^2(E)} \text{if } E \subset \Omega \cup \Gamma,\\
\eta^2 &= \sum_{T \in \mathcal{T}_h} \eta_T^2 + \sum_{E \in \mathcal{E}_h} \eta_E^2.
\end{aligned}$$

8.3 Reissner–Mindlin Plate

The analysis of the non-conforming Reissner–Mindlin plate model Braess (1997); Brezzi and Fortin (1991); Brenner and Scott (1994) due to Arnold and Falk Arnold and Falk (1989) concerns the following (slightly modified) problem for a plane domain Ω with polygonal boundary Γ and a parameter $0 < t < 1$: *Given* $g \in L^2(\Omega)$ *find* $(\phi, w) \in H_0^1(\Omega)^2 \times H_0^1(\Omega)$ *such that, for all* $(\psi, \mu) \in H_0^1(\Omega)^2 \times H_0^1(\Omega)$,

$$(\nabla\phi, \nabla\psi)_{L^2(\Omega)} + t^{-2}(\phi - \nabla w, \psi - \nabla\mu)_{L^2(\Omega)} = (g, \mu)_{L^2(\Omega)}. \tag{43}$$

The discretisation of (43) is based on a regular triangulation $\mathcal{T}$ and conforming and nonconforming P_1 FEM $P_0^1(\mathcal{T})$ and $\mathcal{S}_0^{1,NC}(\mathcal{T})$, respectively; requires additional elementwise cubic bubble functions $\mathcal{B}_3(\mathcal{T})$ and employs the $\mathcal{T}$-piecewise integral mean operator P_0 (i.e., the L^2 projection onto $\mathcal{T}$-piecewise constants): *Find* $(\phi_h, w_h) \in (\mathcal{S}_0^1(\mathcal{T}) \oplus \mathcal{B}_3(\mathcal{T}))^2 \times \mathcal{S}_0^{1,NC}(\mathcal{T})$ *such that, with* $\nabla_{\mathcal{T}}$ *the* $\mathcal{T}$*-piecewise gradient, for all* $(\psi_h, \mu_h) \in (\mathcal{S}_0^1(\mathcal{T}) \oplus \mathcal{B}_3(\mathcal{T}))^2 \times \mathcal{S}_0^{1,NC}(\mathcal{T})$*, there holds*

$$\int_\Omega (\nabla\phi_h \cdot \nabla\psi_h)\,dx + t^{-2}\int_\Omega (P_0\phi_h - \nabla_{\mathcal{T}} w)\cdot(\psi_h - \nabla_{\mathcal{T}}\mu_h)\,dx = \int_\Omega (g\cdot\mu_h)\,dx. \tag{44}$$

With (h, t, g)-independent constants, there holds the a priori error estimate for convex Ω

$$\begin{aligned} |\,\phi - \phi_h\,|_{H^1(\Omega)} + \|r - r_h\|_{L^2(\Omega)} + \|\nabla r - \nabla_{\mathcal{T}} r_h\|_{L^2(\Omega)} + \|\nabla w - \nabla_{\mathcal{T}} w_h\|_{L^2(\Omega)} \\ + \|p - p_h\|_{L^2(\Omega)} + t\,|\,p - p_h\,|_{H^1(\Omega)} \lesssim h\,\|g\|_{L^2(\Omega)}\,. \end{aligned}$$

The additional terms p (resp., r) and their discrete counterparts p_h (resp., r_h) stem from a classical and a discrete Helmholtz decomposition of the exact (resp., discrete) shear ζ (resp., ζ_h) (not required in the numerical scheme) (cf. Arnold and Falk (1989) for proofs):

$$\begin{aligned} \zeta &:= \nabla r + \operatorname{Curl} p = t^{-2}(\nabla w - \phi), \\ \zeta_h &:= \nabla_{\mathcal{T}} r_h + \operatorname{Curl} p_h = t^{-2}(\nabla_{\mathcal{T}} w_h - P_0\phi_h). \end{aligned}$$

This reliable and efficient computable error estimator η is the square root of the sum of all element contributions η_T, $T \in \mathcal{T}$

$$\begin{aligned} \eta_T^2 := \int_T |\phi_h - P_0\phi_h|^2\,dx + h_T^2 \int_T |\zeta_h + \Delta\phi_h|^2\,dx \\ + \min\{1, (h_T/t)^2\} \int_T |\operatorname{Curl}\phi_h|^2\,dx \end{aligned} \tag{45}$$

and of all edge contributions η_E,

$$\eta_E^2 := h_E \int_{E\setminus\Gamma} |[\nabla\phi_h]\cdot\nu_E|^2\,ds + \min\{t^3, t^2 h_E\}\int_E |[\zeta_h]\cdot\tau_E|^2\,ds$$

for an edge E of length h_E with jumps $[\cdot]$ across E and normal and tangential unit vectors ν_E and τ_E. Then

$$\eta := \Big(\sum_{T\in\mathcal{T}}\eta_T^2 + \sum_{E\in\mathcal{E}}\eta_E^2\Big)^{1/2}$$

is a lower and upper bound of the error

$$e_h := |\phi-\phi_h|_{1,2} + \|\nabla w - \nabla_{\mathcal{T}} w_h\|_2 + t\,|p-p_h|_{1,2} + \|p-p_h\|_2\,.$$

in the efficiency and reliability estimate

$$\eta - \text{ h.o.t } \lesssim e_h \lesssim \eta + \text{ h.o.t} \tag{46}$$

The higher-order terms (h.o.t.) in the upper bound in (46), i.e., the reliability of η, are computable terms of r_h such as $(t+h_E)(h_E\int_{E\setminus\Gamma}|[\nabla_{\mathcal{T}} r_h]|^2 ds)^{1/2}$. However, the calculation of r_h requires some extra effort over the solution of (44). The lower bound in (46), i.e., the efficiency of η, holds in a local (elementwise; resp., patchwise) form up to known higher-order terms such as $(t+h_T)h_T\left(\int_T |g|^2\,dx\right)^{1/2}$.

8.4 Goal-Oriented Error Control

The analysis of these lecture notes focused on the estimation of the *energy norm* $\|x-x_h\|_X$. In some applications, there is a given (hence known) linear and bounded *functional* $\rho : X \to \mathbb{R}$ that monitors the error $|\rho(x-x_h)|$ (e.g., the error of an averaged strain or traction over a small but fixed region). To assess the latter quantity, let $A^* : X \to X^*$ be the dual operator of A and let $z \in X$ be the solution to

$$A^* z = \rho.$$

Then it remains to estimate

$$\rho(x-x_h) = (A^*z)(x-x_h) = (A(x-x_h))z = \mathcal{R}es(z).$$

An immediate consequence of this reads

$$|\rho(x-x_h)| \le \|\mathcal{R}es\|_{X^*}\|z\|_X \approx \|\rho\|_{X^*}\|\mathcal{R}es\|_{X^*}. \tag{47}$$

This global estimate (47) is (a) presumably too coarse and (b) does not convey local information of ρ via z. But it indicates that the evaluation of $\mathcal{R}es(z)$ may follow localised arguments from the assessment of $\|\mathcal{R}es\|_{X^*}$ addressed in these lecture notes; see Becker and Rannacher (1996); Ainsworth and Oden (2000); Babuška and Strouboulis (2001) for details and numerical realisations.

A Index of Mathematical Symbols

Basic Notation

$\|E\| = \text{meas}(E)$	Lebesgue measure of $E \subset \mathbb{R}^n$
$\text{card}(\mathcal{E})$	cardinal number of the set $\mathcal{E}$
$u\|_E$	restriction of the function u to the set E
$\text{span}\{v_1, \dots, v_n\}$	vector-space spanned by $v_1, \dots, v_n$
$\dim(X)$	dimension of the vector space X
h.o.t.	abbreviation for "higher order terms"
$\oplus$	direct sum of subspaces
$\forall$	logical symbol "for all"
$\exists$	logical symbol "there exists"
$\exists!$	logical symbol "there exists some unique"

Vectors and Matrices

$\mathbb{R}^{m\times n}$	vector-space of $m \times n$ real valued matrices
$\mathbb{R}^{n\times n}_{\text{sym}}$	$\mathbb{R}^{n\times n}_{\text{sym}} = \{A \in \mathbb{R}^{n\times n} \| A^T = A\}$
$\mathbb{R}^{n\times n}_{\text{skew}}$	$\mathbb{R}^{n\times n}_{\text{skew}} = \{A \in \mathbb{R}^{n\times n} \| A^T = -A\}$

For all $A \in \mathbb{R}^{m\times n}$

a_{ij}	entry of A in the i-th row and j-th column
A^T	transpose of the matrix A

For all $A \in \mathbb{R}^{n\times n}$

$\text{tr}(A)$	trace of A, $tr(A) = \sum_{j=1}^n a_{jj}$
$\det(A)$	determinant of A
$\text{diag}(a_1, a_2, \dots, a_n)$	diagonal matrix $\begin{pmatrix} a_1 & & \\ & \ddots & \\ & & a_n \end{pmatrix}$

$\operatorname{dev} A$	$\operatorname{dev} A := A - \frac{1}{n}\operatorname{tr}(A)$, deviator
$\operatorname{sym}(A)$	$\operatorname{sym}(A) = \frac{1}{2}(A + A^T)$
$\operatorname{skew}(A)$	$\operatorname{skew}(A) = A - \operatorname{sym}(A)$

Differential Operators

∇f	gradient $\nabla f = (\frac{\partial f}{\partial x_1}, \dots, \frac{\partial f}{\partial x_n})^T \in \mathbb{R}^n$ of a function f
$\nabla_{\mathcal{T}}$	elementwise gradient
$\operatorname{div} v$	divergence of v: $\operatorname{div} v = \sum_{j=1}^n \frac{\partial v_j}{\partial x_j} \in \mathbb{R}$ for any $v(x) \in \mathbb{R}^n$, $x \in \mathbb{R}^n$, $\operatorname{div} v = \left(\sum_{j=1}^n \frac{\partial v_{ij}}{\partial x_j}\right)_{1 \le i \le m} \in \mathbb{R}^m$ for any $v(x) \in \mathbb{R}^{m\times n}$, $x \in \mathbb{R}^n$
$\operatorname{curl} v$	$\operatorname{curl} v = \frac{\partial v_2}{\partial x_1} - \frac{\partial v_1}{\partial x_2}$ for any $v(x) \in \mathbb{R}^2$, $x \in \mathbb{R}^2$ $\operatorname{curl} v = (\frac{\partial v_3}{\partial x_2} - \frac{\partial v_2}{\partial x_3}, \frac{\partial v_1}{\partial x_3} - \frac{\partial v_3}{\partial x_1}, \frac{\partial v_2}{\partial x_1} - \frac{\partial v_1}{\partial x_2})$ for any $v(x) \in \mathbb{R}^3$, $x \in \mathbb{R}^3$
$\operatorname{Curl} v$	$\operatorname{Curl} v = (\frac{\partial v}{\partial x_2}, -\frac{\partial v}{\partial x_1})$ for any $v(x) \in \mathbb{R}$, $x \in \mathbb{R}^2$
$\varepsilon(u)$	$\varepsilon(u) = \operatorname{sym}(\nabla u)$

Function Spaces

$L^2(\Omega; X)$	Lebesgue space with values in X, see page 2
$L^2(\Omega)$	abbreviates the spaces $L^2(\Omega;\mathbb{R})$, $L^2(\Omega;\mathbb{R}^n)$, $L^2(\Omega;\mathbb{R}^{n\times n})$
$L^2(\Omega;\mathbb{R}^{n\times n}_{\mathrm{sym}})$	
$L^2(\Omega;\mathbb{R}^{n\times n}_{\mathrm{skew}})$	
$L^2_0(\Omega)$	
$C^{m,\alpha}$	
$H^1(\Omega)$	Sobolev space, see page 2
$H^1_0(\Omega)$	see page 2
$C^\infty_c(\Omega)$	arbitrarily smooth functions with compact support in Ω, see page 2
$H(\operatorname{div},\Omega)$	see page 3
$\ker(A)$	kernel (i.e., nullspace) of linear operator A
$\operatorname{Range}(A)$	range/image of linear operator A
X^*	dual-space of space X

$V^\perp$	$V^\perp := \{x \in X : \forall v \in V \ \langle v, x\rangle_X = 0\}$, $v \subseteq X$ orthogonal complement
$\Vert u\Vert_X$	norm of u in normed space X
$\langle\cdot,\cdot\rangle_X$	scalar product in Hilbert space X
$\perp$	orthogonality in Hilbert spaces
$\mathrm{dist}(X,Y)$	distance, see Remark 2.7 on page 8
γ	trace-operator $\gamma : H^1(\Omega) \to L^2(\partial\Omega)$, $\gamma(u) = u\vert_{\partial\Omega}$
$\langle\cdot,\cdot\rangle_{L^2(\Omega)}$	scalar product in all the spaces $L^2(\Omega)$, $L^2(\Omega;\mathbb{R}^n)$, $L^2(\Omega;\mathbb{R}^{m\times n})$, i.e., $\langle f,g\rangle_{L^2(\Omega)} := \int_\Omega fg\,dx$ for $f,g \in L^2(\Omega)$ or $\int_\Omega f\cdot g\,dx$ respectively $\int_\Omega f : g\,dx$ for $f,g \in L^2(\Omega;\mathbb{R}^n)$ respectively $f,g \in L^2(\Omega;\mathbb{R}^{m\times n})$
$\cdot$	scalar product in $\mathbb{R}^n$
$:$	scalar product in $\mathbb{R}^{m\times n}$
$(\cdot)_+^2$	$(\cdot)_+^2 = (\max\{0,\cdot\})^2$
$\Vert z\Vert_{X\times Y}$	$\Vert z\Vert_{X\times Y} = \sqrt{\langle z,z\rangle_{X\times Y}}$, $z \in X\times Y$, where $\langle (x,y),(u,v)\rangle_{X\times Y} := \langle x,u\rangle_X + \langle y,w\rangle_Y$
$\vert\cdot\vert_{H^1(\Omega)}$	seminorm in $H^1(\Omega)$
$\vert\cdot\vert$	Euclidian norm in $\mathbb{R}, \mathbb{R}^m, \mathbb{R}^{m\times n}$

Mesh-Related Symbols

h_E	$h_E = \mathrm{diam}(E)$, diameter of $E \subset \mathbb{R}^n$
x_E	$x_E = \mathrm{mid}(E)$, midpoint of E
h_T	$h_{\mathcal{T}}\vert_T$, see page 30
$\mathrm{diam}(T)$	$\mathrm{diam}(T) = h_T$
$\mathcal{T}$	triangulation, $T \in \mathcal{T}$ triangle
$\mathcal{E}$	set of edges E of a triangulation $\mathcal{T}$
$\mathcal{E}_\Omega$	set of interior edges, see page 20
$\mathcal{N}$	set of nodes of a triangulation $\mathcal{T}$
ν_E	unit normal vector of the edges E
τ_E	unit tangent vector of the edges E
ω_T	$\omega_T = \bigcup\{T' \in \mathcal{T}\vert\ \overline{T}\cap\overline{T}' \neq \emptyset\}$, patch of the triangle T
ω_E	$\omega_E = \bigcup\{T \in \mathcal{T}\vert\ E \subset \overline{T}\}$, patch of the edge E
ω_z	$\omega_z := \{x \in \Omega : 0 < \phi_z(x)\}$
ψ_E	edge basis function of the lowest Raviart-Thomas element, see page 19
ϕ_z	nodal basis function $\phi_z(y) = \begin{cases} 0 & y \in \mathcal{N}\setminus\{z\} \\ 1 & y = z \in \mathcal{N} \end{cases}$

$[q_h]_E$	jump of q_h across E
$⨍_T f(x)\,dx$	$⨍_T f(x)\,dx := \lvert T\rvert^{-1} \int_T f(x)\,dx$
f_T	$f_T := ⨍_T f(x)\,dx$
$f_{\mathcal{T}}$	$\mathcal{T}$-piecewise constant function: $f_{\mathcal{T}}\vert_T := f_T$
η	see, e.g., page 30, $\eta := \left(\sum_{E\in\mathcal{E}} h_E \lVert[p_h]\rVert^2_{L^2(E)}\right)^{1/2}$
osc	see, e.g., page 30, $\operatorname{osc}(f,\mathcal{T}) := \left(\sum_{T\in\mathcal{T}} h_T^2 \,\lVert f - f_T\rVert_{L^2(T)}\right)^{1/2}$
$\mathcal{R}es$	local definition of the residual, e.g., equation (29) on page 34
$A \lesssim B$	abbreviation for $A \le cB$ with a mesh-size independent, generic constant $c > 0$
$A \approx B$	abbreviation for $A \lesssim B \lesssim A$

Finite Element Spaces

FE	Finite Element
$RT_k(T)$, $RT_k(\mathcal{T})$, RT_k	Raviart-Thomas FEM, see Definition 3.11 on page 15
$P_k(T)$, $P_k(\mathcal{T})$	see Example 3.9 on page 15
$\mathcal{S}^k(\mathcal{T})$, $\mathcal{S}^k_0(\mathcal{T})$	see Example 3.9 on page 15
BDM	Brezzi, Douglas, Marini element, see page 31
$BDFM$	Brezzi, Douglas, Fortin, Marini element, see page 31
B_0	see page 46
BDM_k	see page 46
$BDMS_k$	see page 46
$\mathcal{B}_3(\mathcal{T})$	cubic bubble functions
$S^{1,NC}$, $S^{1,NC}_0$	Crouzeix-Raviart FEM space, see page 44
$B_{k-1}(T)$	see page 46
FEM	finite element method
AFEM	adaptive FEM
MFEM	mixed FEM
NCFEM	non-conforming FEM
AMFEM	adaptive MFEM
ANCFEM	adaptive NCFEM
RT MFEM	Raviart-Thomas MFEM
PEERS	plane elastic elements with reduced symmetries

$\mathcal{R}_k(\partial T)$	functions on ∂T which equal a polynomial of degree at most k on each edge
$\mathcal{P}_0$	$\mathcal{T}$-piecewise integral mean operator, see page 48

Bibliography

M. Ainsworth and J.T. Oden. *A posteriori error estimation in finite element analysis.* Wiley-Interscience [John Wiley & Sons], New York, 2000.

A. Alonso. Error estimators for a mixed method. *Numer. Math.*, 74(4): 385–395, 1996.

D.N. Arnold and R.S. Falk. A uniformly accurate finite element method for the Reissner-Mindlin plate model. *SIAM J. Numer. Anal.*, 26:1276–1290, 1989.

D.N. Arnold and R. Winther. Mixed finite elements for elasticity. *Numer. Math.*, 42:401–419, 2002.

D.N. Arnold, F. Brezzi, and J. Douglas. PEERS: A new mixed finite element for plane elasticity. *Japan J. Appl. Math.*, 1(2):347–367, 1984.

I. Babuška and A.K. Azis. *Survey Lectures on the Mathematical Foundations of the Finite Element Method.* Academic Press.

I. Babuška and A. Miller. A feedback finite element method with a posteriori error estimation: Part i, the finite element method and some properties of the a posteriori estimator. *Comp. Methods Appl. Mech. Engrg.*, 61(1): 1–40, 1987.

I. Babuška and W.C. Rheinboldt. A posteriori error analysis of finite element solutions for one–dimensional problems. *SIAM Journal on Numerical Analysis*, 18:565–589, 1981.

I. Babuška and T. Strouboulis. *The Finite Element Method and its Reliability*, volume xii+802. The Clarendon Press Oxford University Press, 2001.

C. Bahriawati and C. Carstensen. Three Matlab implementations of the lowest-order Raviart-Thomas MFEM with a posteriori error control. *Computational Methods in Applied Mathematics*, 5(4):333–361, 2005.

S. Bartels and C. Carstensen. Each averaging technique yields reliable a posteriori error control in FEM on unstructured grids. II. Higher order FEM. *Math. Comp.*, 71(239):971–994, 2002.

R. Becker and R. Rannacher. A feed-back approach to error control in finite element methods: basic analysis and examples. *East-West J. Numer. Math*, 4:237–264, 1996.

R. Becker and R. Rannacher. An optimal control approach to a posteriori error estimation in finite element methods. *Acta Numerica, Cambridge University Press*, pages 1–102, 2001.

D. Braess. *Finite Elements.* Cambridge University Press, 1997.

D. Braess and R. Verfürth. A posteriori error estimators for the Raviart-Thomas element. *SIAM J. Numer. Anal*, 33:2431–2444, 1996.

D. Braess, C. Carstensen, and B.D. Reddy. Uniform convergence and a posteriori error estimators for the enhanced strain finite element method. *Numer. Math.*, 96:461–479, 2004.

S.C. Brenner and L.R. Scott. *The mathematical theory of finite element methods*, volume 15. Springer Verlag, New York, Texts in Applied Mathematics, 1994.

F. Brezzi and M. Fortin. *Mixed and Hybrid Finite Element Methods.* Springer-Verlag, New York, 1991.

C. Carstensen. A posteriori error estimate for the mixed finite element method. *Math. Comp.*, 66:465–476, 1997.

C. Carstensen. Quasi-interpolation and a a posteriori error analysis in finite element methods. *M2AN*, 33:1187–1202, 1999.

C. Carstensen. A unifying theory of a posteriori finite element error control. *Numer. Math.*, 100:617–637, 2005.

C. Carstensen and G. Dolzmann. A posteriori error estimates for mixed FEM in elasticity. *Numer. Math.*, 81(2):187–209, 1998.

C. Carstensen and S.A. Funken. A posteriori error control in low-order finite element discretisations of incompressible stationary flow problems. *Math. Comp.*, 70(236):1353–1381, 2001a.

C. Carstensen and S.A. Funken. Averaging technique for FE-a posteriori error control in elasticity. part i: Conforming FEM, part ii. λ-independent estimates, part iii: Locking-free conforming FEM. *Comput. Methods Appl. Mech. Engrg.*, 190, 190, 191(8–10):2483–2498, 4663–4675, 861–877, 2001b.

C. Carstensen and S.A. Funken. Constants in Clément-interpolation error and residual based a posteriori error estimates in finite element methods. *East-West J. Numer. Math.*, 8(3):153–175, 2000. ISSN 0928-0200.

C. Carstensen and R. Hoppe. Error reduction and convergence for an adaptive finite element method. *Math. Comp.*, 75(255):1037–1042, 2006.

C. Carstensen and R. Verfürth. Edge residuals dominate a posteriori error estimates for low order finite element methods. *SIAM J. Numer. Anal.*, 36(5):1571–1587, 1999.

C. Carstensen, G. Dolzmann, S.A. Funken, and D.S. Helm. Locking-free adaptive mixed finite element methods in linear elasticity. *Comput. Methods Appl. Mech. Engrg.*, 190(13–14):1701–1718, 2000.

C. Carstensen, S. Bartels, and G. Dolzmann. A posteriori error estimates for nonconforming finite element methods. *Numer. Math.*, 92:233–256, 2002a.

C. Carstensen, S. Bartels, and R. Klose. An experimental survey of a posteriori Courant finite element error control for the Poisson equation. *Adv. Comput. Math.*, 15(1–4):79–106, 2002b.

P. Clément. Approximation by finite element functions using local regularization. *RAIRO Sér. Rouge Anal. Numér.*, 2:77–84, 1975.

E. Dari, R. Durán, and C. Padra. Error estimators for nonconforming finite element approximations of the Stokes problem. *Math. Comp.*, 64(211): 1017–1033, 1995.

E. Dari, R. Durán, C. Padra, and V. Vampa. A posteriori error estimators for nonconforming finite element methods. *RAIRO Modél. Math. Anal. Numér.*, 30(4):385–400, 1996.

K. Eriksson and C. Johnson. Adaptive finite element methods for parabolic problems I. A linear model problem. *SIAM J. Numer. Anal.*, 28:43–77, 1991.

V. Girault and P. A. Raviart. *Finite Element Methods for Navier-Stokes Equations. Theory and algorithms.* Springer Series in Computational Mathematics, 5. Springer- Verlag, Berlin, 1986.

R.H.W. Hoppe and B. Wohlmuth. Element-orientated and edge-orientated local error estimates for nonconforming finite element methods. *Math. Modeling Numer. Anal.*, 30:237–263, 1996.

R. Kouhia and R. Stenberg. A linear nonconforming finite element method for nearly incompressible elasticity and Stokes flow. *Comput. Methods Appl. Mech. Engrg.*, 124(3):195–212, 1995.

L. D. Marini. An inexpensive method for the evaluation of the solution of the lowest order Raviart-Thomas mixed method. *SIAM J. Numer. Anal*, 22:493–496, 1985.

R. Stenberg. A family of mixed finite elements for the elasticity problem. *Numer. Math.*, 53(5):513–538, 1988.

R. Verfürth. *A review of a posteriori error estimation and adaptive mesh-refinement techniques.* Wiley-Teubner, 1996.

R. Verfürth. A posteriori error estimators for the Stokes equations. *Numer. Math.*, 55(3):309–325, 1989.

Fast Solvers for Mixed Finite Element Methods

Susanne C. Brenner*

* Center for Computation and Technology, Johnston Hall, Louisiana State University,
Baton Rouge, LA 70803 USA

Abstract Discretizations of partial differential equations by mixed finite element methods result in large saddle point systems that require efficient solvers. In this chapter we present two classes of iterative methods for such problems: the (preconditioned) Uzawa algorithm and the (preconditioned) minimum residual algorithm. The implementations of these algorithms require either efficient preconditioners for the discrete Laplace operator and/or the efficient solution of the discrete Poisson problem. Towards this end we provide brief introductions to additive Schwarz (domain decomposition/multilevel) preconditioners and multigrid algorithms, after a discussion on block diagonal preconditioners for saddle point problems.

1 Introduction

The discrete problems resulting from mixed finite element methods are saddle point problems. Among the many solution methods for saddle point problems (see Benzi et al., 2005), we will focus on the Uzawa algorithms and the MINRES algorithms.

We consider symmetric saddle point problems of the form

$$\begin{bmatrix} A & B^T \\ B & -C \end{bmatrix} \begin{bmatrix} x \\ y \end{bmatrix} = \begin{bmatrix} f \\ g \end{bmatrix}, \tag{1}$$

where $A \in \mathbb{R}^{n\times n}$ is symmetric positive definite, $B \in \mathbb{R}^{m\times n}$ has full rank m ($m < n$), and $C \in \mathbb{R}^{m\times m}$ is symmetric positive semi-definite. The saddle point problem (1) can be written concisely as

$$\mathcal{A}u = b, \tag{2}$$

where $\mathcal{A}$ is the $\ell \times \ell$ ($\ell = m + n$) matrix

$$\mathcal{A} = \begin{bmatrix} A & B^T \\ B & -C \end{bmatrix}, \quad u = \begin{bmatrix} x \\ y \end{bmatrix} \quad \text{and} \quad b = \begin{bmatrix} f \\ g \end{bmatrix}.$$

Alternatively, we can express (1) in terms of bilinear forms:

$$(Ax, x') + (Bx', y) = (f, x') \qquad \forall x' \in \mathbb{R}^n, \tag{3a}$$

$$(Bx, y) - (Cy, y') = (g, y') \qquad \forall y' \in \mathbb{R}^m, \tag{3b}$$

where $(v, w) = v^T w$ is the Euclidean inner product.

Exercise 1.1. Show that the matrix $\mathcal{A}$ is nonsingular.

The following are some examples of saddle point problems arising from mixed finite element methods, where the standard notation for Sobolev spaces (see Brenner and Scott, 2002) is adopted. We refer the readers to Brezzi and Fortin (1991) for the choices of stable finite element pairs for each of these examples.

Example 1.2. (Poisson Problem)
Find $(\boldsymbol{p}_h, u_h) \in \boldsymbol{Q}_h \times V_h$ such that

$$\begin{aligned} (\boldsymbol{p}_h, \boldsymbol{q})_{L_2(\Omega)} + (\nabla \cdot \boldsymbol{q}, u_h)_{L_2(\Omega)} &= 0 && \forall \boldsymbol{q} \in \boldsymbol{Q}_h, \\ (\nabla \cdot \boldsymbol{p}_h, v)_{L_2(\Omega)} &= (g, v)_{L_2(\Omega)} && \forall v \in V_h, \end{aligned} \tag{4}$$

where $\boldsymbol{Q}_h \times V_h \subset H(\mathrm{div}; \Omega) \times L_2(\Omega)$ is a stable pair of finite element spaces.

Example 1.3. (Stokes Problem)
Find $(\boldsymbol{u}_h, p_h) \in \boldsymbol{V}_h \times Q_h$ such that

$$\begin{aligned} (\nabla \boldsymbol{u}_h, \nabla \boldsymbol{v})_{L_2(\Omega)} + (\nabla \cdot \boldsymbol{v}, p_h)_{L_2(\Omega)} &= (\boldsymbol{f}, \boldsymbol{v})_{L_2(\Omega)} && \forall \boldsymbol{v} \in \boldsymbol{V}_h, \\ (\nabla \cdot \boldsymbol{u}_h, q)_{L_2(\Omega)} &= 0 && \forall q \in Q_h, \end{aligned} \tag{5}$$

where $\boldsymbol{V}_h \times Q_h \subset [H_0^1(\Omega)]^d \times L_2^0(\Omega)$ is a stable pair of finite element spaces.

Example 1.4. (Pure Displacement Elasticity Problem)
Find $(\boldsymbol{u}_h, p_h) \in \boldsymbol{V}_h \times Q_h$ such that

$$\begin{aligned} \mu(\nabla \boldsymbol{u}_h, \nabla \boldsymbol{v})_{L_2(\Omega)} + \quad (\nabla \cdot \boldsymbol{v}, p_h)_{L_2(\Omega)} &= (\boldsymbol{f}, \boldsymbol{v})_{L_2(\Omega)} && \forall \boldsymbol{v} \in \boldsymbol{V}_h, \\ (\nabla \cdot \boldsymbol{u}_h, q)_{L_2(\Omega)} - \frac{1}{\mu + \lambda}(p_h, q)_{L_2(\Omega)} &= 0 && \forall q \in Q_h, \end{aligned} \tag{6}$$

where $\boldsymbol{V}_h \times Q_h \subset [H_0^1(\Omega)]^d \times L_2^0(\Omega)$ is a stable pair of finite element spaces.

The rest of this article is organized as follows. We discuss the Uzawa algorithms in Section 2, the MINRES algorithms in Section 3 and block diagonal preconditioners in Section 4. It follows from the discussions in these sections that good preconditioners/fast solvers for discrete Laplace operators are crucial for fast solvers for saddle point problems. We discuss in the last sections two approaches to the construction of preconditioners/solvers for discrete Laplace operators: additive Schwarz preconditioners in Section 5 and multigrid methods in Section 6.

Remark 1.5. Parallel implementations of many of the algorithms in this article are discussed in Douglas et al. (2003).

2 The Uzawa Algorithms

2.1 The Uzawa Algorithm for (1)

Starting with initial guess (x_0, y_0), the Uzawa algorithm is defined by the following iteration step:

Algorithm 2.1.

$$Ax_{k+1} = f - B^T y_k, \tag{7a}$$

$$y_{k+1} = y_k + \omega(Bx_{k+1} - Cy_k - g). \tag{7b}$$

The iteration for the y-component,

$$y_{k+1} = y_k + \omega\big((BA^{-1}f - g) - (BA^{-1}B^T + C)y_k\big), \tag{8}$$

is precisely Richardson relaxation applied to the SPD system

$$(BA^{-1}B^T + C)y = BA^{-1}f - g, \tag{9}$$

which is obtained from (1) by eliminating x.

Exercise 2.2. Show that (9) can be obtained from (1) by block Gaussian elimination.

The convergence of the y-component is then determined by the well-known result for the Richardson relaxation scheme (cf. Saad, 2003, Section 4.2.1), while the convergence of the x-component follows from the convergence of the y-component.

Theorem 2.3. *Let $M \in \mathbb{R}^{m\times m}$ be SPD. The optimal choice of the parameter ω in the Richardson relaxation scheme*

$$z_{k+1} = z_k + \omega(b - Mz_k) \tag{10}$$

for the system $Mz = b$ is

$$\omega_{\mathrm{opt}} = \frac{2}{\lambda_{\max}(M) + \lambda_{\min}(M)},$$

in which case

$$\|z - z_{k+1}\|_2 \leq \left(\frac{\lambda_{\max}(M) - \lambda_{\min}(M)}{\lambda_{\max}(M) + \lambda_{\min}(M)}\right)\|z - z_k\|_2 = \left(\frac{\kappa_2(M) - 1}{\kappa_2(M) + 1}\right)\|z - z_k\|_2,$$

where

$$\kappa_2(M) = \lambda_{\max}(M)/\lambda_{\min}(M)$$

is the condition number of M with respect to the 2-norm.

Exercise 2.4. Prove Theorem 2.3 using the Spectral Theorem.

2.2 The Preconditioned Uzawa Algorithm for (1)

From Lemma 2.3 we see that the convergence of the Richardson relaxation is slow if $\kappa_2(M) \gg 1$. In such cases the convergence of the Richardson iteration can be improved by employing an SPD preconditioner $\mathcal{P} \in \mathbb{R}^{m\times m}$, where the iteration step (10) is replaced by

$$z_{k+1} = z_k + \omega\mathcal{P}^{-1}(b - Mz_k). \tag{11}$$

Exercise 2.5. Show that $\mathcal{P}^{-1}M$ is SPD with respect to the inner product $(\cdot,\cdot)_{\mathcal{P}} = \sqrt{(\mathcal{P}\cdot,\cdot)}$.

In view of Exercise 2.5, the analysis of the Richardson relaxation scheme carries over to the preconditioned Richardson relaxation (11) and Lemma 2.3 remains valid if we replace $\|\cdot\|_2 = \sqrt{(\cdot,\cdot)}$ by $\|\cdot\|_{\mathcal{P}} = \sqrt{(\mathcal{P}\cdot,\cdot)}$. The convergence rate of (11) is faster than that of (10) provided that

$$\frac{\lambda_{\max}(P^{-1}M)}{\lambda_{\min}(P^{-1}M)} < \frac{\lambda_{\max}(M)}{\lambda_{\min}(M)}. \tag{12}$$

The discussion above leads to the following preconditioned Uzawa algorithm for (1):

Algorithm 2.6.

$$Ax_{k+1} = f - B^T y_k, \tag{13a}$$

$$y_{k+1} = y_k + \omega \mathcal{P}^{-1}(Bx_{k+1} - Cy_k - g), \tag{13b}$$

where $\mathcal{P} \in \mathbb{R}^{m\times m}$ is SPD.

2.3 Examples

The application of the Uzawa algorithms in Section 2.1 and Section 2.2 to Example 1.2–Example 1.4 requires good preconditioners and/or fast solvers for discrete Laplace operators.

Example 2.7. In the saddle point formulation (4) of the Poisson problem, we have $A = M_h$, where M_h is a mass matrix representing the L_2 inner product; $B = B_h$, where the matrix B_h represents a discrete divergence operator; and $C = 0$. In this case the matrix $BA^{-1}B^T + C = B_h M_h^{-1} B_h^T$ represents a nonconforming discrete Laplace operator and $\kappa_2(B_h M_h^{-1} B_h^T) \approx h^{-2}$, where h is the mesh size of the underlying triangulation. We need a good preconditioner for the nonconforming discrete Laplace operator in order to apply Algorithm 2.6.

Example 2.8. For the saddle point problem (5) of the Stokes equations, we have $A = A_h$, where the matrix A_h represents a discrete Laplace operator; $B = B_h$, where the matrix B_h represents a discrete divergence operator; and $C = 0$. In this case the matrix $BA^{-1}B^T + C = B_h A_h^{-1} B_h^T$ is well conditioned and we can solve (5) by Algorithm 2.1. However the execution of A_h^{-1} requires a fast solver for a discrete Poisson problem.

Example 2.9. For the saddle point problem (6), we have $A = \mu A_h$, where the matrix A_h represents a discrete Laplace operator; $B = B_h$, the matrix representing a discrete divergence operator; and $C = C_h = 1/(\mu + \lambda)M_h$, where M_h is a mass matrix representing the L_2 inner product. In this case the matrix $BA^{-1}B^T + C = B_h A_h^{-1} B_h^T + C_h$ is well-conditioned and we can solve (6) by Algorithm 2.1. Again the evaluation of A^{-1} requires a fast solver for a discrete Poisson problem.

2.4 The Augmented Lagrangian Method

In the case where $C = 0$, the saddle point problem (1) can be solved by the following iteration:

$$(A + \gamma B^T B)x_{k+1} = f - B^T y_k + \gamma B^T g, \tag{14a}$$

$$y_{k+1} = y_k + \rho(Bx_{k+1} - g). \tag{14b}$$

The iteration (14) is precisely the Uzawa algorithm applied to the system

$$\begin{bmatrix} A+\gamma B^T B & B^T \\ B & 0 \end{bmatrix} \begin{bmatrix} x \\ y \end{bmatrix} = \begin{bmatrix} f+\gamma B^T g \\ g \end{bmatrix},$$

where $\gamma > 0$ is a penalty parameter.

It can be shown that the iterative method (14) converges for appropriately chosen ρ and γ. In particular, it converges if $\rho = \gamma > 0$ is sufficiently large. Details can be found in Fortin and Glowinski (1983) and Brenner and Scott (2002).

3 The MINRES Algorithms

Given an initial guess u_0, the Minimum Residual (MINRES) algorithm computes

$$u_k = \underset{v \in u_0 + K_k(\mathcal{A}, r_0)}{\operatorname{argmin}} \|b - \mathcal{A}v\|_2 \tag{15}$$

as an approximate solution to (2), where $K_k(\mathcal{A}, r_0)$ is the k-dimensional Krylov subspace generated by the initial residual $r_0 = b - \mathcal{A}u_0$, i.e.,

$$K_k(\mathcal{A}, r_0) = \langle r_0, \mathcal{A}r_0, \ldots, \mathcal{A}^{k-1}r_0 \rangle. \tag{16}$$

Exercise 3.1. Show that u_s is the exact solution of (2), where $\ell = m + n$ and s equals the dimension of $K_\ell(\mathcal{A}, r_0)$.

3.1 Error Estimates for the MINRES Algorithm

Although MINRES is a direct method for (2), it is usually treated as an iterative method when $\ell \gg 1$. An error estimate for $\|b - \mathcal{A}u_k\|_2$ can be obtained through the theory of polynomial approximation.

Exercise 3.2. Show that $\{b - \mathcal{A}v : v \in K_k(\mathcal{A}, r_0)\} = \{p(\mathcal{A})r_0 : p \in \mathbb{P}_k \text{ and } p(0) = 1\}$, where $\mathbb{P}_k$ is the space of polynomials of degree $\leq k$.

It follows from (15) and Exercise 3.2 that

$$\|b - \mathcal{A}u_k\| = \min_{p \in \mathbb{P}_k, p(0)=1} \|p(\mathcal{A})r_0\|_2. \tag{17}$$

Exercise 3.3. Use the Spectral Theorem to show that $\|p(A)\| = \max_{\lambda \in \sigma(\mathcal{A})} |p(\lambda)|$, where $\sigma(\mathcal{A})$ is the spectrum of $\mathcal{A}$.

Combining (17) and Exercise 3.3, we see that

$$\|b - \mathcal{A}u_k\|_2 \leq \min_{p \in \mathbb{P}_k, p(0)=1} \max_{\lambda \in \sigma(\mathcal{A})} |p(\lambda)|,$$

which implies the following estimate:

$$\|b - \mathcal{A}u_k\|_2 \le \big(\max_{\lambda \in J} |p(\lambda)|\big) \|r_0\| \qquad \text{for any } p \in \mathbb{P}_k \text{ such that } p(0) = 1, \tag{18}$$

where $J = [-\beta, -\alpha] \cup [\alpha, \beta]$,

$$\alpha = \min_{\lambda \in \sigma(\mathcal{A})} |\lambda| \quad \text{and} \quad \beta = \max_{\lambda \in \sigma(\mathcal{A})} |\lambda|.$$

For $k = 2j$ or $k = 2j+1$ ($j \geq 0$), we can obtain a concrete error estimate for the MINRES algorithm by choosing the polynomial p in (18) to be

$$p_k(t) = \frac{T_j(q(t))}{T_j(q(0))}, \tag{19}$$

where T_j is the j-th Chebyshev polynomial defined by the relation $T_j(\cos\theta) = \cos(j\theta)$ for $0 \le \theta \le \pi$ and $q(t)$ is the quadratic polynomial defined by

$$q(t) = \frac{\beta^2 + \alpha^2 - 2t^2}{\beta^2 - \alpha^2}$$

that maps J onto $[-1, 1]$.

Exercise 3.4. Show that (cf. Greenbaum, 1997, Section 3.1)

$$\max_{\lambda \in J} |p_k(\lambda)| = \frac{2}{\left(\frac{\beta+\alpha}{\beta-\alpha}\right)^j + \left(\frac{\beta-\alpha}{\beta+\alpha}\right)^j}.$$

The estimate (18) with $p = p_k$ defined by (19) together with Exercise 3.4 yields the following result for the MINRES algorithm.

Theorem 3.5. *Let u_k ($k \geq 1$) be the outputs of the MINRES algorithm with initial guess u_0. Then we have*

$$\|r_k\|_2 = \min_{p \in \mathbb{P}_k,\, p(0)=1} \|p(\mathcal{A})r_0\|_2, \tag{20}$$

where $r_0 = b - \mathcal{A}u_0$ is the initial residual and $r_k = b - \mathcal{A}u_k$ is the k-th residual. A concrete estimate is

$$\|r_k\|_2 \le 2\left(\frac{\kappa_2(\mathcal{A}) - 1}{\kappa_2(\mathcal{A}) + 1}\right)^{[k/2]} \|r_0\|_2, \tag{21}$$

where $[k/2]$ is the integer part of $k/2$ and

$$\kappa_2(\mathcal{A}) = \frac{\max_{\lambda \in \sigma(\mathcal{A})} |\lambda|}{\min_{\lambda \in \sigma(\mathcal{A})} |\lambda|}$$

is the condition number of $\mathcal{A}$ with respect to the 2-norm.

3.2 Implementation of the MINRES Algorithm

In order to solve the least squares problem (15), we construct an orthonormal basis $v_1, \ldots, v_k$ for the Krylov space $K_k(\mathcal{A}, r_0)$, by applying the Gram-Schmidt process to the vectors $v_1, \ldots, v_{k-1}$ and $\mathcal{A}v_{k-1}$. By the symmetry of $\mathcal{A}$, the orthonormal vectors satisfy a three term recurrence ($v_0 = 0 = v_{\ell+1}$):

$$\mathcal{A}v_j = \gamma_j v_{j-1} + \delta_j v_j + \gamma_{j+1} v_{j+1} \qquad \text{for} \quad 1 \le j \le \ell. \tag{22}$$

The vectors v_j can be generated efficiently through the relation (22), resulting in the well-known Lanczos process.

Exercise 3.6. Verify the recurrence relation (22).

Let $V_k \in \mathbb{R}^{\ell \times k}$ be the matrix whose columns are $v_1, \ldots, v_k$. The recurrence relation (22) can be written as

$$\mathcal{A}V_k = V_{k+1} T_{k+1}, \tag{23}$$

where $T_{k+1} \in \mathbb{R}^{(k+1)\times k}$ is the tridiagonal matrix with subdiagonal $(\gamma_2, \ldots, \gamma_{k+1})$, diagonal $(\delta_1, \ldots, \delta_k)$ and superdiagonal $(\gamma_2, \ldots, \gamma_k)$.

Writing $v = u_0 + V_k x$ for $x \in \mathbb{R}^k$ in (15), we find by (23)

$$b - \mathcal{A}v = b - \mathcal{A}u_0 - \mathcal{A}V_k x = r_0 - V_{k+1} T_{k+1} x = V_{k+1}\big(\gamma_1 e_1^{(k+1)} - T_{k+1} x\big), \tag{24}$$

where $e_1^{(k+1)}$ is the canonical basis vector $(1, 0, \ldots, 0)^T \in \mathbb{R}^{k+1}$. It follows from (24) and the orthonormality of $v_1, \ldots, v_{k+1}$ that the solution of (15) is $u_k = u_0 + V_k x_k$, where $x_k \in \mathbb{R}^k$ satisfies

$$x_k = \operatorname*{argmin}_{x \in \mathbb{R}^k} \|\gamma_1 e_1^{(k+1)} - T_{k+1} x\|_2. \tag{25}$$

Since $T_{k+1} \in \mathbb{R}^{(k+1)\times k}$ is tridiagonal, we can compute the QR factorization of T_{k+1} by Givens rotations and then solve the least squares problem (25). Furthermore, since T_{k+1} has only three more nonzero components than T_k and they appear in the last column, the QR factorizations and hence the solutions of the least squares problems can be generated by simple updating. The resulting algorithm (cf. Elman et al., 2005, Section 2.4) is stated below.

Algorithm 3.7. (MINRES for (2))

$v_0 = 0, w_0 = 0, w_1 = 0$

Choose u_0, compute $v_1 = b - \mathcal{A}u_0$, set $\gamma_1 = \|v_1\|_2$

Set $\eta = \gamma_1$, $s_0 = s_1 = 0$, $c_0 = c_1 = 1$

for $j = 1$ until convergence do

  % Lanczos process

    $v_j \leftarrow v_j/\gamma_j$

    $\delta_j = (\mathcal{A}v_j, v_j)$

    $v_{j+1} = \mathcal{A}v_j - \delta_j v_j - \gamma_j v_{j-1}$

    $\gamma_{j+1} = \|v_{j+1}\|_2$

  % Update QR factorization

    $\alpha_0 = c_j\delta_j - c_{j-1}s_j\gamma_j$

    $\alpha_1 = \sqrt{\alpha_0^2 + \gamma_{j+1}^2}$

    $\alpha_2 = s_j\delta_j + c_{j-1}c_j\gamma_j$

    $\alpha_3 = s_{j-1}\gamma_j$

  % Givens rotation

    $c_{j+1} = \alpha_0/\alpha_1$, $s_{j+1} = \gamma_{j+1}/\alpha_1$

  % Least squares solution

    $w_{j+1} = (v_j - \alpha_3 w_{j-1} - \alpha_2 w_j)/\alpha_1$

    $u_j = u_{j-1} + c_{j+1}\eta w_{j+1}$

  % Update right-hand side

    $\eta = -s_{j+1}\eta$

    Test for convergence ($|\eta| = \|b - \mathcal{A}u_j\|_2$)

enddo

Exercise 3.8. The Lanczos process breaks down if $\gamma_j = 0$. Show that in this case u_{j-1} is the exact solution of (2).

3.3 The Preconditioned MINRES Algorithms

As indicated by (21), the convergence of the MINRES algorithm is slow when $\kappa_2(\mathcal{A}) \gg 1$, in which case we should use the following preconditioned MINRES algorithm (cf. Elman et al., 2005, Section 6.1).

Algorithm 3.9. (Preconditioned MINRES for (2))

Let $\mathcal{P} \in \mathbb{R}^{(m+n)\times(m+n)}$ be SPD.

$v_0 = 0, w_0 = 0, w_1 = 0$

Choose u_0, compute $v_1 = b - \mathcal{A}u_0$

Solve $\mathcal{P}z_1 = v_1$, set $\gamma_1 = \sqrt{(z_1, v_1)}$

Set $\eta = \gamma_1$, $s_0 = s_1 = 0$, $c_0 = c_1 = 1$
for $j = 1$ until convergence do
  % Lanczos process
    $z_j \leftarrow z_j/\gamma_j$
    $\delta_j = (\mathcal{A}z_j, z_j)$
    $v_{j+1} = Az_j - (\delta_j/\gamma_j)v_j - (\gamma_j/\gamma_{j-1})v_{j-1}$
    Solve $\mathcal{P}z_{j+1} = v_{j+1}$
    $\gamma_{j+1} = \sqrt{(z_{j+1}, v_{j+1})}$
  % Update QR factorization
    $\alpha_0 = c_j\delta_j - c_{j-1}s_j\gamma_j$
    $\alpha_1 = \sqrt{\alpha_0^2 + \gamma_{j+1}^2}$
    $\alpha_2 = s_j\delta_j + c_{j-1}c_j\gamma_j$
    $\alpha_3 = s_{j-1}\gamma_j$
  % Givens rotation
    $c_{j+1} = \alpha_0/\alpha_1$, $s_{j+1} = \gamma_{j+1}/\alpha_1$
  % Least squares solution
    $w_{j+1} = (z_j - \alpha_3 w_{j-1} - \alpha_2 w_j)/\alpha_1$
    $u_j = u_{j-1} + c_{j+1}\eta w_{j+1}$
  % Update right-hand side
    $\eta = -s_{j+1}\eta$
    Test for convergence ($|\eta| = \|\mathcal{P}^{-1}(b - \mathcal{A}u_j\|_2$)
enddo

Algorithm 3.9 computes

$$u_k = \underset{v \in u_0 + K_k(\mathcal{P}^{-1}\mathcal{A}, \tilde{r}_0)}{\text{arumin}} \|\mathcal{P}^{-1}(b - \mathcal{A}v)\|_{\mathcal{P}},$$

where $\tilde{r}_0 = \mathcal{P}^{-1}(b - \mathcal{A}u_0) = \mathcal{P}^{-1}r_0$ is the preconditioned initial residual, $K_k(\mathcal{P}^{-1}\mathcal{A}, \tilde{r}_0)$ is the Krylov subspace spanned by $\{\tilde{r}_0, \mathcal{P}^{-1}\mathcal{A}\tilde{r}_0, \ldots, (\mathcal{P}^{-1}\mathcal{A})^{k-1}\tilde{r}_0\}$, and $\|\cdot\|_{\mathcal{P}} = \sqrt{(\mathcal{P}\cdot,\cdot)}$.

Using the fact that $\mathcal{P}^{-1}\mathcal{A}$ is symmetric with respect to the inner product $(\cdot,\cdot)_{\mathcal{P}} = (\mathcal{P}\cdot,\cdot)$, the analysis in Section 3.1 carries over and we have the following result for Algorithm 3.9.

Theorem 3.10. *Let u_k $(k \geq 1)$ be the output of the preconditioned MINRES algorithm with initial guess u_0. Then we have*

$$\|\tilde{r}_k\|_{\mathcal{P}} = \min_{p \in \mathbb{P}_k, p(0)=1} \|p(\mathcal{P}^{-1}\mathcal{A})\tilde{r}_0\|_{\mathcal{P}}, \tag{26}$$

where $\tilde{r}_0 = \mathcal{P}^{-1}(b - \mathcal{A}u_0)$ *is the initial (preconditioned) residual and* $\tilde{r}_k = \mathcal{P}^{-1}(b - \mathcal{A}u_k)$ *is the k-th (preconditioned) residual. A concrete estimate is*

$$\|\tilde{r}_k\|_{\mathcal{P}} \leq 2\left(\frac{\kappa_{\mathcal{P}}(\mathcal{P}^{-1}\mathcal{A}) - 1}{\kappa_{\mathcal{P}}(\mathcal{P}^{-1}\mathcal{A}) + 1}\right)^{[k/2]} \|\tilde{r}_0\|_{\mathcal{P}}, \tag{27}$$

where $[k/2]$ *is the integer part of* $k/2$ *and*

$$\kappa_{\mathcal{P}}(\mathcal{P}^{-1}\mathcal{A}) = \frac{\max_{\lambda\in\sigma(\mathcal{P}^{-1}\mathcal{A})}|\lambda|}{\min_{\lambda\in\sigma(\mathcal{P}^{-1}\mathcal{A})}|\lambda|}$$

is the condition number of $\mathcal{A}$ *with respect to the* $\mathcal{P}$*-norm.*

Exercise 3.11. Show that $\|\tilde{r}_k\|_{\mathcal{P}} = \|r_k\|_{\mathcal{P}^{-1}}$, where $\|\cdot\|_{\mathcal{P}^{-1}} = \sqrt{(\mathcal{P}^{-1}\cdot,\cdot)}$.

4 Block Diagonal Preconditioners for Saddle Point Problems

In this section we consider block diagonal preconditioners for the saddle point problem (2) and refer the readers to Benzi et al. (2005) and the references therein for a discussion of block triangular preconditioners.

4.1 The Case $C = 0$

We first consider the preconditioner

$$\mathcal{P}_* = \begin{bmatrix} A & 0 \\ 0 & -S \end{bmatrix}, \tag{28}$$

where $S = -BA^{-1}B^T$ is the Schur complement matrix for

$$\mathcal{A} = \begin{bmatrix} A & B^T \\ B & 0 \end{bmatrix}.$$

A simple calculation gives

$$\mathcal{P}_*^{-1}\mathcal{A} = \begin{bmatrix} I_{m\times m} & A^{-1}B^T \\ -S^{-1}B & 0 \end{bmatrix}, \tag{29}$$

where $I_{m\times m}$ is the identity matrix in $\mathbb{R}^{m\times m}$. Note that

$$(-S^{-1}B)(A^{-1}B^T) = -S^{-1}(BA^{-1}B^T) = -S^{-1}(-S) = I. \tag{30}$$

Exercise 4.1. Let a be a nonzero number. Show that the 2×2 matrix $M = \begin{bmatrix} 1 & 1/a \\ a & 0 \end{bmatrix}$ satisfies the relation $M^2 - M - I = 0$.

The following lemma generalizes Exercise 4.1.

Lemma 4.2. *Let* $M = \begin{bmatrix} I_{m\times m} & F \\ E & 0 \end{bmatrix}$*, where* $E \in \mathbb{R}^{n\times m}$*,* $F \in \mathbb{R}^{m\times n}$ *and* $EF = I_{n\times n}$*. Then* M *satisfies the relation*

$$M(M-I)(M^2 - M - I) = 0. \tag{31}$$

Proof. Let I be the $(m+n)\times(m+n)$ identity matrix. We have

$$M(M-I) = \begin{bmatrix} FE & 0 \\ 0 & I_{n\times n} \end{bmatrix}. \tag{32}$$

Since $EF = I_{n\times n}$, it follows from (32) that

$$[M(M-I)]^2 = M(M-I),$$

which implies (31). □

Let $M = \mathcal{P}_*^{-1}\mathcal{A}$. In view of (30), Lemma 4.2 and that M is nonsingular, we have

$$(M-I)(M^2 - M - I) = 0. \tag{33}$$

Exercise 4.3. Show that $(\lambda - 1)(\lambda^2 - \lambda - 1)$ is the minimal polynomial of $\mathcal{P}_*^{-1}\mathcal{A}$ (Murphy et al., 2000).

It follows from Exercise 4.3 that $\mathcal{P}_*^{-1}\mathcal{A}$ has only three distinct eigenvalues: 1, $(1+\sqrt{5})/2$ and $(1-\sqrt{5})/2$. In particular, by taking $p(\lambda) = (\lambda-1)(\lambda^2-\lambda-1)$, we see from (18) that Algorithm 3.9 converges in three steps if we take the preconditioner $\mathcal{P}$ to be $\mathcal{P}_*$.

However, the interests concerning the preconditioner $\mathcal{P}_*$ are more of a theoretical nature since the evaluation of $\mathcal{P}_*^{-1}u$ is too expensive. In practice, one would usually replaces $\mathcal{P}_*$ by

$$\mathcal{P} = \begin{bmatrix} \hat{A} & 0 \\ 0 & -\hat{S} \end{bmatrix}, \tag{34}$$

where $\hat{A}$ (respectively $\hat{S}$) is a good approximation of A (respectively S) such that $\hat{A}^{-1}v$ (respectively $\hat{S}^{-1}w$) can be efficiently computed. The effect of such a preconditioner is given in the following theorem (cf. Toselli and Widlund, 2004, Section 9.2).

Theorem 4.4. *Let $S = -BAB^T$ be the Schur complement matrix for the saddle point system*

$$\mathcal{A} = \begin{bmatrix} A & B^T \\ B & 0 \end{bmatrix}.$$

Let $\hat{A} \in \mathbb{R}^{n\times n}$ and $-\hat{S} \in \mathbb{R}^{m\times m}$ be SPD matrices such that

$$a_1(Ax, x) \le (\hat{A}x, x) \le a_2(Ax, x) \qquad \forall\, x \in \mathbb{R}^n, \tag{35}$$

$$c_1(-Sy, y) \le (-\hat{S}y, y) \le c_2(-Sy, y) \qquad \forall\, y \in \mathbb{R}^m, \tag{36}$$

then we have

$$\kappa_{\mathcal{P}}(\mathcal{P}^{-1}\mathcal{A}) \le \left(\frac{\sqrt{5}+1}{\sqrt{5}-1}\right) \frac{\max\{a_2, c_2\}}{\min\{a_1, c_1\}}, \tag{37}$$

where the block diagonal preconditioner $\mathcal{P}$ is given by (34).

Exercise 4.5. Prove Theorem 4.4.

4.2 The Case $C \neq 0$

We will consider block diagonal preconditioners for the system

$$\mathcal{A}_t = \begin{bmatrix} A & B^T \\ B & -t^2C \end{bmatrix} \qquad (0 \le t \le 1) \tag{38}$$

and begin with the preconditioner

$$\mathcal{P}_* = \begin{bmatrix} A & 0 \\ 0 & C \end{bmatrix}. \tag{39}$$

Exercise 4.6. Show that if $\lambda \neq 1$ is an eigenvalue of $\mathcal{P}_*^{-1}\mathcal{A}_t$, then $(\lambda - 1)(\lambda + t^2)$ is an eigenvalue of $C^{-1}(BA^{-1}B^T)$.

It follows from Exercise 4.6 that if λ is an eigenvalue of $\mathcal{P}_*^{-1}\mathcal{A}_t$, then either $\lambda = 1$ or $(\lambda-1)(\lambda+t^2) = \lambda'$ for some eigenvalue λ' of $C^{-1}(BA^{-1}B^T)$, i.e.,

$$\lambda = \frac{1-t^2}{2} \pm \sqrt{\lambda' + \frac{(1+t^2)^2}{4}}. \tag{40}$$

Exercise 4.7. Let λ be an eigenvalue of $\mathcal{P}_*^{-1}\mathcal{A}_t$. Show that

$$\min\left(1, -\frac{1}{2}+\sqrt{\lambda_{\min}(C^{-1}BA^{-1}B^T) + \frac{1}{4}}\right) \le |\lambda| \le \frac{1}{2}+\sqrt{\lambda_{\max}(C^{-1}BA^{-1}B^T) + \frac{1}{4}}.$$

The estimates in Exercise 4.7 imply

$$\kappa_{\mathcal{P}_*}(\mathcal{P}_*^{-1}\mathcal{A}_t) = \frac{\max_{\lambda\in\sigma(\mathcal{P}_*^{-1}\mathcal{A}_t)}|\lambda|}{\min_{\lambda\in\sigma(\mathcal{P}_*^{-1}\mathcal{A}_t)}|\lambda|} \le \frac{\gamma_1}{\gamma_2},$$

where

$$\gamma_1 = \frac{1}{2} + \sqrt{\lambda_{\max}(C^{-1}BA^{-1}B^T) + \frac{1}{4}}, \tag{41}$$

$$\gamma_2 = \min\left(1, -\frac{1}{2} + \sqrt{\lambda_{\min}(C^{-1}BA^{-1}B^T) + \frac{1}{4}}\right). \tag{42}$$

Again the evaluation of $\mathcal{P}_*^{-1}u$ is too expensive and in practice it is replaced by the preconditioner

$$\mathcal{P} = \begin{bmatrix} \hat{A} & 0 \\ 0 & \hat{C} \end{bmatrix}, \tag{43}$$

where $\hat{A}$ (respectively $\hat{C}$) is a good approximation of A (respectively C) such that $\hat{A}^{-1}v$ (respectively $\hat{C}^{-1}w$) can be efficiently computed. The effect of such a preconditioner is given in the following lemma (cf. Toselli and Widlund, 2004, Section 9.4).

Theorem 4.8. *Let $\hat{A} \in \mathbb{R}^{n\times n}$ and $\hat{C} \in \mathbb{R}^{m\times m}$ be SPD matrices such that*

$$a_1(Ax,x) \le (\hat{A}x,x) \le a_2(Ax,x) \qquad \forall x \in \mathbb{R}^n, \tag{44}$$

$$c_1(Cy,y) \le (\hat{C}y,y) \le c_2(Cy,y) \qquad \forall y \in \mathbb{R}^m, \tag{45}$$

then we have

$$\kappa_{\mathcal{P}}(\mathcal{P}^{-1}\mathcal{A}_t) \le \left(\frac{\gamma_1}{\gamma_2}\right)\frac{\max\{a_2,c_2\}}{\min\{a_1,c_1\}} \qquad 0 \le t \le 1, \tag{46}$$

where $\mathcal{A}_t$ is the matrix given by (38), γ_1 and γ_2 are given in (41) and (42), and $\mathcal{P}$ is the preconditioner in (43).

Exercise 4.9. Prove Theorem 4.8.

4.3 Examples

The results in Section 4.1 can be applied to the saddle point problems in Example 1.2 and Example 1.3, while the results in Section 4.2 can be applied to the saddle point problem in Example 1.4. In the following examples we will follow the notation introduced in Example 2.7–Example 2.9.

Example 4.10. For the saddle point problem (4), the matrix $A = M_h$ is well-conditioned and we can take $\hat{A} = c\,h^2 I_{m\times m}$. The Schur complement matrix $-S = B_h M_h^{-1} B_h^T$ represents a nonconforming discrete Laplace operator and we can take $-\hat{S}$ to be a good preconditioner for this nonconforming discrete Laplace operator.

Example 4.11. For the saddle point problem (5), the matrix $A = A_h$ represents a discrete Laplace operator and we can take $\hat{A}$ to be a good preconditioner for this discrete Laplace operator. The Schur complement operator $S = -B_h A_h B_h^T$ is well-conditioned and we can take $\hat{S} = c\,I_{n\times n}$.

Example 4.12. For the saddle point problem (6), the matrix $A = A_h$ represents a discrete Laplace operator and we need a good preconditioner $\hat{A}$ for this discrete Laplace operator. The matrix $C = M_h$ is well conditioned and we can take $\hat{C} = c\,h^2 I_{n\times n}$.

5 Additive Schwarz Preconditioners

We discuss additive Schwarz preconditioners for discrete Laplace operators in this section. This class of preconditioners is particularly suitable for parallel implementation. We begin with the abstract theory which is followed by the descriptions of several concrete preconditioners. Detailed discussions of these preconditioners can be found in Quarteroni and Valli (1999), Brenner and Scott (2002), and Toselli and Widlund (2004).

5.1 The Abstract Theory of Additive Schwarz Preconditioners

An additive Schwarz preconditioner $\mathcal{B} \in \mathbb{R}^{n\times n}$ for a SPD matrix $A \in \mathbb{R}^{n\times n}$ is defined by the following expression:

$$\mathcal{B}^{-1} = \sum_{j=0}^{J} I_j A_j^{-1} I_j^T, \tag{47}$$

where $A_j \in \mathbb{R}^{n_j\times n_j}$ is SPD, $I_j \in \mathbb{R}^{n\times n_j}$ and given any $v \in \mathbb{R}^n$,

$$v = \sum_{j=0}^{J} I_j v_j \quad \text{where} \quad v_j \in \mathbb{R}^{n_j} \quad \text{for} \quad 0 \le j \le J. \tag{48}$$

Exercise 5.1. Let $v \in \mathbb{R}^n$ and $I_j^T v = 0$ for $0 \le j \le J$. Show that (48) implies $v = 0$.

It is clear from (47) that $\mathcal{B}^{-1}$ is symmetric positive semi-definite. Furthermore, $(\mathcal{B}^{-1}v, v) = 0$ implies $\sum_{j=0}^{J}(A_j^{-1}I_j^T v, I_j^T v) = 0$ and hence $I_j^T v = 0$ for $0 \leq j \leq J$. It follows from Exercise 5.1 that $v = 0$ and hence $\mathcal{B}^{-1}$ (and therefore $\mathcal{B}$) is SPD.

The abstract theory of additive Schwarz preconditioners is based on a formula for $(\mathcal{B}v, v)$. Let $v = \sum_{j=0}^{J} I_j v_j$. We have

$$
\begin{aligned}
(\mathcal{B}v, v) &= (\mathcal{B}v, \sum_{j=0}^{J} I_j v_j) \\
&= \sum_{j=0}^{J}(A_j^{-1}I_j^T \mathcal{B}v, A_j v_j) \\
&\leq \sum_{j=0}^{J}(A_j^{-1}I_j^T \mathcal{B}v, A_j A_j^{-1} I_j^T \mathcal{B}v)^{1/2}(v_j, A_j v_j)^{1/2} \\
&= \sum_{j=0}^{J}(I_j A_j^{-1} I_j^T \mathcal{B}v, \mathcal{B}v)^{1/2}(v_j, A_j v_j)^{1/2} \\
&\leq \sqrt{\sum_{j=0}^{J}(I_j A_j^{-1} I_j^T \mathcal{B}v, \mathcal{B}v)}\sqrt{\sum_{j=0}^{J}(v_j, A_j v_j)} \\
&= \sqrt{(v, \mathcal{B}v)}\sqrt{\sum_{j=0}^{J}(v_j, A_j v_j)},
\end{aligned}
$$

which implies that

$$
(\mathcal{B}v, v) \leq \sum_{j=0}^{J}(A_j v_j, v_j) \qquad \text{whenever} \quad v = \sum_{j=0}^{J} I_j v_j. \tag{49}
$$

Exercise 5.2. Let $v \in R^n$ and $v_j = A_j^{-1} I_j^T \mathcal{B}v$. Show that $v = \sum_{j=0}^{J} I_j v_j$ and

$$
(Bv, v) = \sum_{j=0}^{J}(A_j v_j, v_j).
$$

We summarize the discussion so far by the next lemma which follows immediately from (49) and Exercise 5.2.

Lemma 5.3. *Under condition* (48), *the operator* $\mathcal{B}$ *is SPD and*

$$(\mathcal{B}v, v) = \min_{v=\sum_{j=0}^{J} I_j v_j} \sum_{j=0}^{J} (A_j v_j, v_j). \tag{50}$$

Corollary 5.4. *Under condition* (48), *the eigenvalues of* $\mathcal{B}^{-1}A$ *are positive and*

$$\lambda_{\max}(\mathcal{B}^{-1}A) = \max_{v\in\mathbb{R}^n\setminus\{0\}} \frac{(Av, v)}{\min_{v=\sum_{j=0}^{J} I_j v_j} \sum_{j=0}^{J} (A_j v_j, v_j)}, \tag{51}$$

$$\lambda_{\min}(\mathcal{B}^{-1}A) = \min_{v\in\mathbb{R}^n\setminus\{0\}} \frac{(Av, v)}{\min_{v=\sum_{j=0}^{J} I_j v_j} \sum_{j=0}^{J} (A_j v_j, v_j)}. \tag{52}$$

Proof. Since $\mathcal{B}^{-1}A$ is SPD with respect to the inner product $(\cdot,\cdot)_{\mathcal{B}} = (\mathcal{B}\cdot,\cdot)$, the formulas (51) and (52) follow from well-known properties of Rayleigh quotients and (50). □

We have given a flexible version of the theory of additive Schwarz preconditioners initiated by M. Dryja and O. Widlund. When applied to a specific preconditioner, the algebraic formulas (51) and (52) must be combined with analytical tools to derive bounds for the condition number of the preconditioned system.

5.2 The Two-Level Additive Schwarz Preconditioner

Let the triangulation $\mathcal{T}_h$ be a refinement of a coarse triangulation $\mathcal{T}_H$ of the domain Ω. The domain Ω is divided into an *overlapping* decomposition $\Omega_1, \ldots, \Omega_J$, where each subdomain is the union of elements from $\mathcal{T}_h$. A typical decomposition is depicted in Figure 1 where δ represents the amount of overlap among the subdomains.

The stiffness matrix for the Poisson problem (with homogeneous Dirichlet boundary condition) associated with $\mathcal{T}_h$ (respectively $\mathcal{T}_H$) is denoted by $A_h \in \mathbb{R}^{n_h\times n_h}$ (respectively $A_H \in \mathbb{R}^{n_H\times n_H}$). $A_j \in \mathbb{R}^{n_j\times n_j}$, $(1 \le j \le J)$, is the stiffness matrix for the Poisson problem (with homogeneous Dirichlet boundary condition) on the subdomain Ω_j with respect to $\mathcal{T}_h$. $I_j \in \mathbb{R}^{n_h\times n_j}$ is the Boolean matrix that maps the degrees of freedom on Ω_j to the corresponding degrees of freedom on Ω and $I_H \in \mathbb{R}^{n_h\times n_H}$ is the interpolation

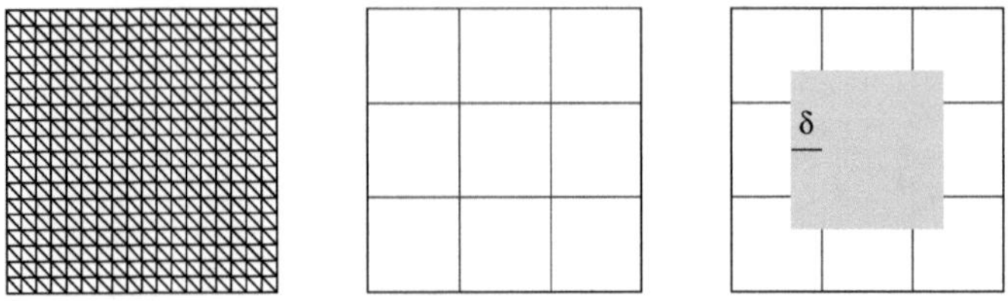

Figure 1. An Overlapping Domain Decomposition

(prolongation) matrix that represents the degrees of freedom associated with $\mathcal{T}_H$ in terms of the degrees of freedom associated with $\mathcal{T}_h$. In the discussion below we will use the same notation to denote a finite element function and the coordinate vector that represents it.

The two-level additive Schwarz preconditioner introduced in Dryja and Widlund (1987) is defined by

$$\mathcal{B}_{TL}^{-1} = I_H A_H^{-1} I_H^T + \sum_{j=1}^{J} I_j A_j^{-1} I_j^T .$$

Exercise 5.5. Establish the condition (48) for the two-level additive Schwarz preconditioner.

We begin the analysis of $\mathcal{B}_{TL}^{-1}$ by establishing an upper bound for $\lambda_{\max}(\mathcal{B}_{TL}^{-1} A_h)$. Suppose we can find $\alpha > 0$ such that

$$(A_h v, v) \le \alpha \Big[(A_H v_H, v_H) + \sum_{j=1}^{J} (A_j v_j, v_j) \Big] \tag{53}$$

whenever $v \in \mathbb{R}^{n_h}$, $v_H \in \mathbb{R}^{n_H}$, $v_j \in \mathbb{R}^{n_j}$ and $v_h = I_H v_H + \sum_{j=1}^{J} I_j v_j$, then it follows from (51) that

$$\lambda_{\max}(\mathcal{B}_{TL}^{-1} A_h) \le \alpha. \tag{54}$$

Exercise 5.6. Show that (54) follows from (53) and (51).

The estimate (53) can be translated into

$$\|\nabla v_h\|_{L_2(\Omega)}^2 \le \alpha \Big[\|\nabla v_H\|_{L_2(\Omega)}^2 + \sum_{j=1}^{J} \|\nabla v_j\|_{L_2(\Omega)}^2 \Big], \tag{55}$$

whenever $v_h = v_H + \sum_{j=1}^{J} v_j$ as finite element functions, and we can establish (55) as follows:

$$\begin{aligned}
\|\nabla v_h\|^2_{L_2(\Omega)} &= \|\nabla v_H + \sum_{j=1}^{J} \nabla v_j\|^2_{L_2(\Omega)} \\
&\le \Big(\|\nabla v_H\|_{L_2(\Omega)} + \|\sum_{j=1}^{J} \nabla v_j\|_{L_2(\Omega)}\Big)^2 \\
&\le 2\Big(\|\nabla v_H\|^2_{L_2(\Omega)} + \|\sum_{j=1}^{J} \nabla v_j\|^2_{L_2(\Omega)}\Big) \le 2\Big(\|\nabla v_H\|^2_{L_2(\Omega)} + \alpha_* \sum_{j=1}^{J} \|\nabla v_j\|^2_{L_2(\Omega)}\Big),
\end{aligned}$$

where $\alpha_* > 0$ depends only on the maximum number of subdomains that can overlap with any one of the subdomains. Therefore (55) and hence (54) hold for $\alpha = 2\max(1, \alpha_*)$.

Exercise 5.7. Prove the estimate

$$\|\sum_{j=1}^{J} \nabla v_j\|^2_{L_2(\Omega)} \le \alpha_* \sum_{j=1}^{J} \|\nabla v_j\|^2_{L_2(\Omega)}$$

using the fact that $\int_\Omega \nabla v_j \nabla v_k \, dx = 0$ if Ω_j and Ω_k are disjoint.

We now turn to a lower bound for $\lambda_{\min}(\mathcal{B}_{TL}^{-1} A_h)$. Suppose we can find $\beta > 0$ so that for any $v_h \in \mathbb{R}^{n_h}$, there exists a decomposition $v = I_H v_H + \sum_{j=1}^{J} I_j v_j$ such that

$$(A_H v_H, v_H) + \sum_{j=1}^{J} (A_j v_j, v_j) \le \beta (A_h v_h, v_h), \tag{56}$$

then it follows from (52) that

$$\lambda_{\min}(\mathcal{B}_{TL}^{-1} A_h) \ge \frac{1}{\beta}. \tag{57}$$

Exercise 5.8. Show that (57) follows from (56) and (52).

Given $v_h \in \mathbb{R}^{n_h}$, the decomposition satisfying (56) is constructed as follows. First we construct a partition of unity $\theta_1, \ldots, \theta_J$ with the following properties:

(i) $\theta_j \in C^\infty(\bar{\Omega})$ and $\theta_j \ge 0$,

(ii) $\theta_j = 0$ outside Ω_j,
(iii) $\sum_{j=1}^{J} \theta_j = 1$ on $\bar{\Omega}$,
(iv) $|\nabla\theta_j| \le C/\delta$.

Then we define $v_H = Q_H v_h$ and $v_j = \Pi_h\big[\theta_j(v_h - v_H)\big]$, where Q_H is the L_2-orthogonal projection onto the finite element space associated with $\mathcal{T}_H$ and Π_h is the nodal interpolation operator onto the finite element space associated with $\mathcal{T}_h$.

The following estimate is valid for the decomposition $v_h = v_H + \sum_{j=1}^{J} v_j$:

$$\|\nabla v_H\|_{L_2(\Omega)}^2 + \sum_{j=1}^{J} \|\nabla v_j\|_{L_2(\Omega)}^2 \le C_{**}\Big(1 + \frac{H}{\delta}\Big)\|\nabla v\|_{L_2(\Omega)}^2, \tag{58}$$

which translates into (56) with $\beta = C_{**}\Big(1 + H/\delta\Big)$. The proof of (58) involves the estimate (Bramble and Xu, 1991)

$$\|v_h - Q_H v_h\|_{L_2(\Omega)} + H\|\nabla(Q_H v_h)\|_{L_2(\Omega)} \le C'\|\nabla v_h\|_{L_2(\Omega)}.$$

Details can be found in Brenner and Scott (2002) and Toselli and Widlund (2004).

Combining the estimates for $\lambda_{\max}(\mathcal{B}_{TL}^{-1}A_h)$ and $\lambda_{\min}(\mathcal{B}_{TL}^{-1}A_h)$, we have

$$\frac{\lambda_{\max}(\mathcal{B}_{TL}^{-1}A_h)}{\lambda_{\min}(\mathcal{B}_{TL}^{-1}A_h)} \le \gamma\left(1 + \frac{H}{\delta}\right), \tag{59}$$

where the positive constant γ depends on the shape regularity of $\mathcal{T}_h$, $\mathcal{T}_H$ and $\Omega_1, \ldots, \Omega_J$, and the maximum number of subdomains that can overlap any one of the subdomains. The estimate (59) is sharp (cf. Brenner, 2000).

Remark 5.9. The two-level additive Schwarz preconditioner is scalable in the sense that the constant γ in (59) is independent of J. This is due to the fact that global communication among the subdomains is provided through the coarse space. Without the coarse space, we have a so-called one-level additive Schwarz preconditioner

$$\mathcal{B}_{OL}^{-1} = \sum_{j=1}^{J} I_j A_j^{-1} I_j^t$$

and it can be proved (cf. Brenner (2006)) that

$$\frac{\lambda_{\max}(\mathcal{B}_{OL}^{-1}A_h)}{\lambda_{\min}(\mathcal{B}_{OL}^{-1}A_h)} \ge C_* J,$$

i.e., it is not scalable.

Remark 5.10. The exact solves A_H^{-1} and A_j^{-1} can be replaced by inexact solves $\tilde{A}_H^{-1}$ and $\tilde{A}_j^{-1}$ in (5.2) as long as

$$(A_H v_H, v_H) \approx (\tilde{A}_H v_H, v_H) \quad \text{and} \quad (A_j v_j, v_j) \approx (\tilde{A}_j v_j, v_j).$$

For example, one can use multigrid solves instead of exact solves.

5.3 The Neumann-Neumann Preconditioner

The matrices $A_h \in \mathbb{R}^{n_h \times n_h}$ $A_H \in \mathbb{R}^{n_H \times n_H}$ and $I_H \in \mathbb{R}^{n_h \times n_H}$ have the same meanings as those in Subsection 5.2. $M_h \in \mathbb{R}^{n_h \times n_h}$ is the mass matrix. The domain Ω is divided into *nonoverlapping* subdomains $\Omega_1, \ldots, \Omega_J$ and $\Gamma_j = \partial\Omega_j \setminus \Omega$ is the part of the boundary of Ω_j that is disjoint from the boundary of Ω.

We denote by

$$\begin{bmatrix} A_{II}^{(j)} & A_{IB}^{(j)} \\ A_{BI}^{(j)} & A_{BB}^{(j)} \end{bmatrix}$$

the sub-assembled stiffness matrix on $\bar{\Omega}_j$ corresponding to the degrees of freedom $(x_I^{(j)}, x_B^{(j)})$ in the interior of Ω_j and on Γ_j. $A_{II}^{(j)} \in \mathbb{R}^{n_j \times n_j}$ $(1 \leq j \leq J)$ is a SPD matrix, where n_j is the number of interior degrees of freedom for Ω_j.

Similarly,

$$\begin{bmatrix} M_{II}^{(j)} & M_{IB}^{(j)} \\ M_{BI}^{(j)} & M_{BB}^{(j)} \end{bmatrix}$$

is the sub-assembled mass matrix on $\bar{\Omega}_j$.

For each j, the matrix $\hat{S}_j \in \mathbb{R}^{m_j \times m_j}$ (m_j is the number of degrees of freedom on Γ_j) defined by

$$\hat{S}_j = [A_{BB}+H^{-2}M_{BB}]-[A_{BI}+H^{-2}M_{BI}][A_{II}+H^{-2}M_{II}]^{-1}[A_{IB}+H^{-2}M_{IB}]$$

is the Schur complement matrix of

$$\begin{bmatrix} A_{II}^{(j)} & A_{IB}^{(j)} \\ A_{BI}^{(j)} & A_{BB}^{(j)} \end{bmatrix} + \frac{1}{H^2} \begin{bmatrix} M_{II}^{(j)} & M_{IB}^{(j)} \\ M_{BI}^{(j)} & M_{BB}^{(j)} \end{bmatrix}$$

resulted from eliminating the interior degrees of freedom.

Remark 5.11. $\hat{S}_j$ is the Schur complement matrix for the bilinear form

$$\hat{D}_j(v_j, w_j) = \int_{\Omega_j} \nabla v_j \cdot \nabla w_j \, dx + \frac{1}{H^2} \int_{\Omega_j} v_j w_j \, dx,$$

where v_j and w_j are finite element functions defined on $\bar{\Omega}_j$. Note that $\hat{D}_j$ is positive-definite for all subdomains while the bilinear form

$$D_j(v_j, w_j) = \int_{\Omega_j} \nabla v_j \cdot \nabla w_j \, dx$$

is only positive-semidefinite for floating subdomains.

The Neumann-Neumann preconditioner (Dryja and Widlund, 1995) is defined by

$$\mathcal{B}_{NN}^{-1} = I_H A_H^{-1} I_H^T + \sum_{j=1}^{J} E_j (A_{II}^{(j)})^{-1} E_j^T + \sum_{j=1}^{J} (I_j D_j) \hat{S}_j^{-1} (I_j D_j)^T,$$

where $E_j \in \mathbb{R}^{n_h \times n_j}$ is the Boolean matrix that maps the interior degrees of freedom of Ω_j to the corresponding global degrees of freedom and $D_j \in \mathbb{R}^{m_j \times m_j}$ is a diagonal scaling matrix such that $D_j(k,k)^{-1}$ is the number of subdomains whose closures contain the k-th node on Γ_j.

Below we outline the analysis of the Neumann-Neumann preconditioner. Details can be found in Brenner and Scott (2002) and Toselli and Widlund (2004).

There exists a positive constant c_*, independent of h, H and J such that

$$\lambda_{\min}(\mathcal{B}_{NN}^{-1} A_h) \geq c_*. \tag{60}$$

The derivation of (60) involves discrete harmonic functions on the subdomains Ω_j, which are finite element functions orthogonal to the finite element functions that vanish on $\partial\Omega_j$, with respect to the bilinear form $D(\cdot,\cdot)$. A discrete harmonic function v_j has minimum energy among all all finite element functions w_j that agree with v_j on $\partial\Omega_j$, i.e.,

$$\|\nabla v_j\|_{L_2(\Omega_j)} \leq \|\nabla w_j\|_{L_2(\Omega_j)} \quad \text{if} \quad w_j = v_j \quad \text{on} \quad \partial\Omega_j. \tag{61}$$

The lower bound (60) follows easily from (52) and (61).

In the other direction we have the estimate

$$\lambda_{\max}(\mathcal{B}_{NN}^{-1} A_h) \leq C_* \Big(1 + \ln \frac{H}{h}\Big)^2, \tag{62}$$

where the positive constant C_* is independent of h, H and J. The estimate (62) is proved by using (51) and the following discrete Sobolev inequality:

$$\|v_j\|_{L_\infty(\partial\Omega_j)} \leq C\Big(1 + \ln \frac{H}{h}\Big) \|v_j\|_{H^{1/2}(\partial\Omega_j)}, \tag{63}$$

which holds for any finite element function v_j on $\bar{\Omega}_j$.

Combining (60) and (62), we have

$$\frac{\lambda_{\max}(\mathcal{B}_{NN}^{-1}A_h)}{\lambda_{\min}(\mathcal{B}_{NN}^{-1}A_h)} \leq C\left(1+\ln\frac{H}{h}\right)^2, \tag{64}$$

where the positive constant C depends only on the shape regularity of $\mathcal{T}_h$ and $\mathcal{T}_H$.

Remark 5.12. The estimate (64) is sharp (Brenner and Sung, 2000).

5.4 The Bramble-Pasciak-Xu Preconditioner

The Bramble-Pasciak-Xu preconditioner (Bramble et al., 1990) is a multilevel preconditioner. Let $\mathcal{T}_1, \ldots, \mathcal{T}_J$ be triangulations of Ω obtained from the triangulation $\mathcal{T}_0$ by uniform subdivision, V_j be the finite element space associated with $\mathcal{T}_j$, and $A_j \in \mathbb{R}^{n_j \times n_j}$ be the stiffness matrix for the Poisson problem with homogeneous Dirichlet boundary condition associated with the triangulation $\mathcal{T}_j$. A_J is the matrix to be preconditioned.

The BPX preconditioner is defined by

$$\mathcal{B}_{BPX}^{-1} = \sum_{j=0}^{J} I_j I_j^T,$$

where $I_j \in \mathbb{R}^{n_J \times n_j}$ is the interpolation (prolongation) matrix that represents the degrees of freedom associated with $\mathcal{T}_j$ in terms of the degrees of freedom associated with $\mathcal{T}_J$.

The upper bound

$$\lambda_{\max}(\mathcal{B}_{BPX}^{-1}A_J) \leq C_*, \tag{65}$$

where $C_* > 0$ is independent of h and J, can be derived using (51) and the following strengthened Cauchy-Schwarz inequality (Xu, 1992):

$$\int_\Omega \nabla v_j \cdot \nabla v_k \, dx \leq C 2^{-|j-k|/2}(h_j^{-1}\|v_j\|_{L_2(\Omega)})(h_k^{-1}\|v_k\|_{L_2(\Omega)}), \tag{66}$$

where v_j (respectively v_k) is a finite element function associated with $\mathcal{T}_i$ (respectively $\mathcal{T}_j$).

In the other direction the lower bound

$$\lambda_{\min}(\mathcal{B}_{BPX}^{-1}A_J) \geq c_*, \tag{67}$$

where $c_* > 0$ is independent of h and J, can be derived using (52) and the following estimate from the theory of computational scales (Bramble et al.,

2000):

$$\|\nabla v\|^2_{L_2(\Omega)} \approx \sum_{j=1}^{J} h_j^{-2} \|\nabla(Q_j v - Q_{j-1} v)\|^2_{L_2(\Omega)} \tag{68}$$

for any $v \in V_J$, where Q_j is the L_2-orthogonal projection operator onto V_j.

Combining (65) and (67), we have the following optimal condition number estimate:

$$\frac{\lambda_{\max}(\mathcal{B}_{BPX}^{-1} A_J)}{\lambda_{\min}(\mathcal{B}_{BPX}^{-1} A_J)} \leq C \tag{69}$$

where the positive constant C depends on Ω and the shape regularity of $\mathcal{T}_0$.

Details for the estimates (65) and (67) can be found in Bramble et al. (1990), Xu (1992) and Brenner and Scott (2002).

5.5 Other Domain Decomposition Preconditioners

We have discussed two domain decomposition preconditioners in this Section. There are of course many other popular domain decomposition algorithms such as the Bramble-Pasciak-Schatz preconditioner (Bramble et al., 1986), the balancing domain decomposition method of Mandel (Mandel, 1993), the finite element tearing and interconnecting method of Farhat (Farhat and Roux, 1991; Farhat et al., 2000), and the balancing domain decomposition method with constraints of Dohrmann (Dohrmann, 2003).

6 Multigrid Methods

We discuss multigrid algorithms for discrete Poisson problems in this Section. Other aspects of multigrid methods can be found in Hackbusch (1985), Wesseling (1992), Bramble and Zhang (2000), Briggs et al. (2001) and Trottenberg et al. (2001).

Let $\mathcal{T}_0$ be an initial triangulation of Ω and $\mathcal{T}_1, \mathcal{T}_2, \ldots$ be triangulations obtained from $\mathcal{T}_0$ by uniform subdivision. $A_k \in \mathbb{R}^{n_k \times n_k}$ is the stiffness matrix for the Poisson problem with homogeneous Dirichlet boundary condition associated with $\mathcal{T}_k$ and $I^k_{k-1} \in \mathbb{R}^{n_k \times n_{k-1}}$ is the interpolation (prolongation) matrix that represents the degrees of freedom associated with $\mathcal{T}_{k-1}$ in terms of the degrees of freedom associated with $\mathcal{T}_k$. The k-th level discrete problem can therefore be written as

$$A_k x = b. \tag{70}$$

6.1 Ideas Behind Multigrid Methods

The SPD system (70) can be solved by an iterative method (Jacobi, Gauss-Seidel, SOR, etc.). After m iterations (with initial guess x_0), we obtain an approximate solution x_m. Then the error e_m satisfy the residual equation

$$A_k e_m = r_m, \tag{71}$$

where the residual error $r_m = b - A_k x_m$.

If we can solve (71) exactly, we can recover the exact solution of (70) by the relation

$$x = x_m + (x - x_m) = x_m + e_m. \tag{72}$$

However, (71) has the same complexity as (70). So we would not solve (71) exactly. Instead, we can solve (71) approximately to obtain an approximation e'_m of e_m, and then, hopefully (cf. (72)), the correction

$$x' = x_m + e'_m \tag{73}$$

will give an approximation of x that is better than the original approximation x_m.

In the multigrid setting there is a natural way to find an approximate solution of (71), namely, by solving the coarse grid residual equation

$$A_{k-1} e_{k-1} = r_{k-1}, \tag{74}$$

where r_{k-1} is some sort of projection of r_m in the coarse space. If (74) is solved exactly, then we can correct the approximation x_m by

$$x_{m+1} = x_m + I_{k-1}^k e_{k-1}. \tag{75}$$

This is known as a *two-grid method.*

But of course (74) does not have to be solved exactly. If we apply the idea of the two-grid method to solve (74) approximately using a coarser grid, and so on, then we have a multigrid algorithm.

By solving the residual equation on the coarse grid, the computational cost is reduced. Therefore multigrid algorithms have optimal complexity in the sense that the costs of these algorithms are proportional to the number of unknowns. Furthermore, since smoothing is carried out on all levels, the errors of different frequencies are damped out uniformly. In other words, multigrid algorithms can overcome the ill-conditioning of the discrete system (70).

6.2 Multigrid Algorithms

Algorithm 6.1. (k-th Level V-cycle Algorithm for (70))

$k = 0$: Output $= A_0^{-1}b$ (direct solve).

$k \geq 1$: Output obtained in three steps.

Pre-Smoothing:

- Apply m steps of a classical iteration scheme (Richardson, Gauss-Seidel) with initial guess x_0 to (70) to obtain an approximate solution x_m.

Coarse Grid Correction

- Transfer the residual of x_m to the coarse grid as

$$\tilde{r} = (I_{k-1}^k)^T (b - A_k x_m).$$

- Apply the $(k-1)$-st level algorithm to the $(k-1)$-st level residual equation

$$A_{k-1}\tilde{e} = \tilde{r}$$

  with initial guess 0 to obtain an approximate solution e_{k-1}.

- Compute the correction

$$z_{m+1} = z_m + I_{k-1}^k e_{k-1}.$$

Post-Smoothing

- Apply m steps of a classical iteration scheme (Richardson, Gauss-Seidel) with initial guess x_{m+1} to (70) to obtain an approximate solution x_{2m+1}.

Output $= x_{2m+1}$.

Algorithm 6.2. (k-th Level W-cycle Algorithm) In the coarse grid correction step applied the $(k-1)$-st level algorithm twice.

Algorithm 6.3. (k-th Level F-Cycle Algorithm) In the coarse grid correction step applied the $(k-1)$-st level algorithm once followed by a $(k-1)$-st level V-cycle algorithm

The scheduling diagrams for the V-cycle, W-cycle and F-cycle algorithms are given in Figure 2, Figure 3 and Figure 4, where the solid dots represents smoothing.

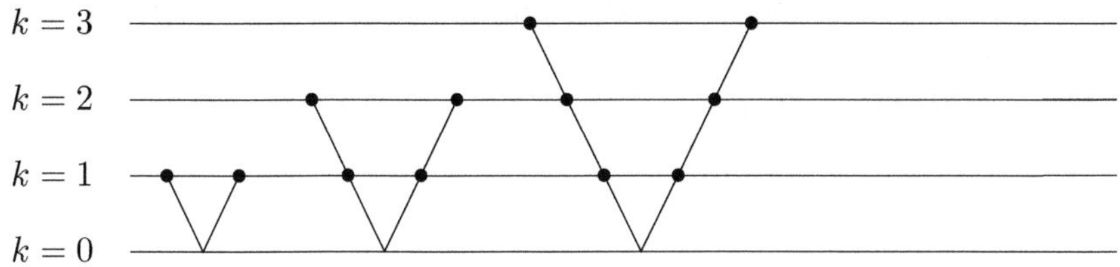

Figure 2. Scheduling Diagram of the V-cycle Algorithm

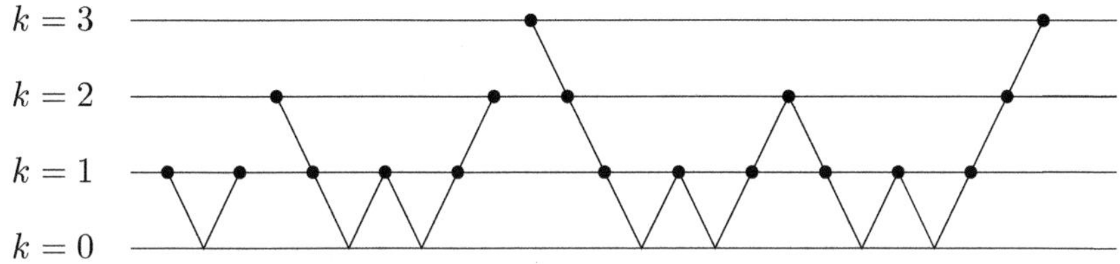

Figure 3. Scheduling Diagram of the W-cycle Algorithm

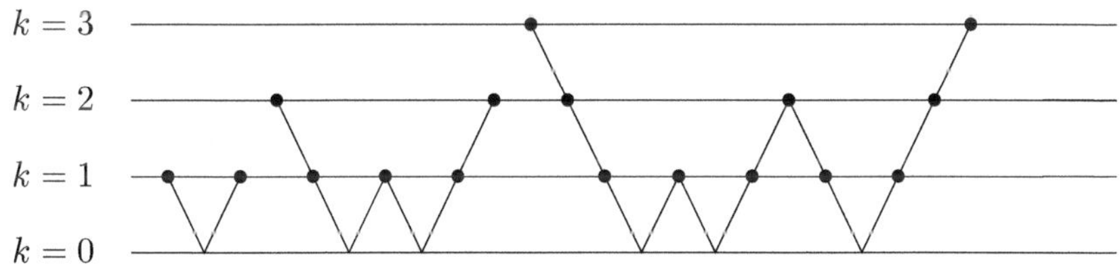

Figure 4. Scheduling Diagram of the F-cycle Algorithm

The V-cycle, W-cycle and F-cycle algorithms are optimal order algorithms in the sense that the cost w_k of executing the k-th level algorithm satisfies the estimate

$$w_k = O(n_k), \tag{76}$$

where n_k is the number of unknowns at the k-th level.

Exercise 6.4. Prove (76) under the assumptions that I_j and A_j are sparse matrices that have only a few nonzero components on each row and each column, and that the cost for executing one smoothing step at the j-th level is of order $O(n_j)$.

Two of the key ingredients for the analysis of multigrid algorithms appear already in the analysis of the two-grid algorithm, where the residual equation

$A_{k-1}\tilde{e} = \tilde{r}$ is solved exactly. Let R_k be the error propagation matrix for one smoothing step, Id_k be the $n_k \times n_k$ identity matrix, and $P_k^{k-1} \in \mathbb{R}^{n_{k-1}\times n_k}$ be the transpose of the matrix $I_{k-1}^k \in \mathbb{R}^{n_k \times n_{k-1}}$ with respect to the inner products $(A_k\cdot,\cdot)$ and $(A_{k-1}\cdot,\cdot)$. Then the final error $\hat{e}_{2m+1}$ of the two-grid algorithm with m pre-smoothing steps and m post-smoothing steps is related to the initial error $\hat{e}_0$ by

$$\hat{e}_{2m+1} = R_k^m(Id_k - I_{k-1}^k P_k^{k-1})R_k^m \hat{e}_0. \tag{77}$$

Exercise 6.5. Establish the relation (77).

It is clear from (77) that the analysis of two-grid algorithms (and hence multigrid algorithms) involves estimates for the matrices R_k^m and $Id_k - I_{k-1}^k P_k^{k-1}$. The estimate for the matrix R_k^m is known as the *smoothing property*, which involves only the Spectral Theorem and calculus. The estimate for the matrix $Id_k - I_{k-1}^k P_k^{k-1}$ is known as the *approximation property*, which involves duality arguments and elliptic regularity estimates.

The following relation between the error propagation matrices E_k and E_{k-1} for V-cycle and W-cycle algorithms on two consecutive levels can be established recursively:

$$E_k = R_k^m(Id_k - I_{k-1}^k P_k^{k-1} + I_{k-1}^k E_{k-1}^p P_k^{k-1})R_k^m, \tag{78}$$

where $p = 1$ (respectively $p = 2$) for the V-cycle (respectively W-cycle) algorithm.

Exercise 6.6. Prove the recursive relation (78).

Once the smoothing property and the approximation property are established, the analysis of the multigrid algorithm proceeds either by induction in the case of the W-cycle algorithm (Bank and Dupont, 1981) or by combining the multiplicative theory (Bramble and Zhang, 2000) and the additive theory (Brenner, 2002) in the case of the V-cycle algorithm. The analysis of the F-cycle algorithm then follows by a perturbation argument.

The following estimate is valid for conforming finite element discretizations :

$$\|x - x_{2m+1}\|_{A_k} \le \frac{C}{C + m^\alpha}\|x - x_0\|_{A_k}, \tag{79}$$

where x_{2m+1} is the output of the V-cycle algorithm, the W-cycle algorithm or the F-cycle algorithm with m pre-smoothing and m post-smoothing steps, the positive constant C is independent of k and m, α is the index of elliptic regularity and $\|\cdot\|_{A_k} = \sqrt{(A_k\cdot,\cdot)}$.

In particular, the estimate (79) indicates that the V-cycle, W-cycle and F-cycle algorithms converge with one smoothing step. On the other hand, when the underlying finite element space is nonconforming (cf. Brenner, 1999, 2004), the estimate (79) must be replaced by

$$\|x - x_{2m+1}\|_{A_k} \le Cm^{-\alpha}\|x - x_0\|_{A_k}, \tag{80}$$

where the positive constant C is independent of m and k, provided $m > m_*$ for a large enough integer m_* that is also independent of k.

Algorithm 6.7. (k-th Level Variable V-cycle Algorithm) In this V-cycle algorithm the number of smoothing steps can depend on k.

Let the numbers of smoothing steps $m_1, m_2, \ldots, m_k$ in the variable V-cycle algorithm satisfy the relation

$$\beta_1 m_j \le m_{j-1} \le \beta_2 m_j \qquad \text{for} \quad 0 \le j \le k, \tag{81}$$

where $1 < \beta_1 \le \beta_2$. Then it holds (Bramble and Zhang, 2000) that

$$\frac{\lambda_{\max}(\mathcal{B}_k^{-1} A_k)}{\lambda_{\min}(\mathcal{B}_k^{-1} A_k)} \le C, \tag{82}$$

where $\mathcal{B}_k^{-1} b =$ the output of the variable V-cycle algorithm for (70) with initial guess 0 and the positive constant C is independent of k. In other words, the variable V-cycle algorithm provides an optimal preconditioner for the discrete Laplace operator.

6.3 The Full Multigrid Method

Consider the Poisson problem of finding $u \in H_0^1(\Omega)$ such that

$$\int_\Omega \nabla u \cdot \nabla v \, dx = \int_\Omega f v \, dx \qquad \forall\, v \in H_0^1(\Omega). \tag{83}$$

The k-th level discrete problem is

$$A_k u_k = f_k, \tag{84}$$

where $f_k \in \mathbb{R}^{n_k}$ satisfies

$$(f_k, v_k) = \int_\Omega f \, v_k \, dx$$

for all finite element functions v_k associated with $\mathcal{T}_k$.

The idea behind the full multigrid method is that one should use the approximate solution of (84) obtained by a multigrid algorithm as the initial guess for the $(k+1)$-st level problem. The resulting algorithm is a multilevel nested iteration of multigrid algorithms.

Algorithm 6.8. (Full Multigrid Method for (83))
$k = 0$: Output $= A_0^{-1} f_0$ (direct solve).
$k \geq 1$: Output obtained recursively as follows.

- $u_0^k = I_{k-1}^k u_{k-1}$
- apply a k-th level multigrid algorithm r-times, with initial guess u_0^k, to obtain the approximate solution u_k of (84)

It can be shown (Brenner and Scott, 2002) that, for r sufficiently large, the outputs u_k are quasi-optimal approximations of u and the cost of the full multigrid method is $O(n_k)$.

Remark 6.9. If the k-th level multigrid algorithm in Algorithm 6.8 is replaced by a k-th level smoothing scheme (without a coarse grid correction), then the resulting algorithm is known as a cascadic multigrid method (Bornemann and Deuflhard, 1996).

6.4 Other Multigrid Methods

We have discussed geometric multigrid methods where the finite element spaces are associated with meshes obtained through geometric refinement of a coarsest mesh. One can also start with the finite element space associated with a fine mesh and created a hierarchy of coarser spaces, using heuristic coarsening strategies. This approach leads to algebraic multigrid methods (Ruge and Stüben, 1987; Briggs et al., 2001; Stüben, 2001) that are usually used as blackbox solvers.

Bibliography

R.E. Bank and T.F. Dupont. An optimal order process for solving finite element equations. *Math. Comp.*, 36:35–51, 1981.

M. Benzi, G.H. Golub, and J. Liesen. Numerical solution of saddle point problems. *Acta Numerica*, 14:1–137, 2005.

F.A. Bornemann and P. Deuflhard. The cascadic multigrid method for elliptic problems. *Numer. Math.*, 75:135–152, 1996.

J.H. Bramble, J.E. Pasciak, and A.H. Schatz. The construction of preconditioners for elliptic problems by substructuring, I. *Math. Comp.*, 47: 103–134, 1986.

J.H. Bramble, J.E. Pasciak, and P.S. Vassilevski. Computational scales of Sobolev norms with application to preconditioning. *Math. Comp.*, 69: 463–480, 2000.

J.H. Bramble, J.E. Pasciak, and J. Xu. Parallel multilevel preconditioners. *Math. Comp.*, 55:1–22, 1990.

J.H. Bramble and J. Xu. Some estimates for a weighted L^2 projection. *Math. Comp.*, 56:463–476, 1991.

J.H. Bramble and X. Zhang. The Analysis of Multigrid Methods. In P.G. Ciarlet and J.L. Lions, editors, *Handbook of Numerical Analysis, VII*, pages 173–415. North-Holland, Amsterdam, 2000.

S.C. Brenner. Convergence of nonconforming multigrid methods without full elliptic regularity. *Math. Comp.*, 68:25–53, 1999.

S.C. Brenner. Lower bounds for two-level additive Schwarz preconditioners with small overlap. *SIAM J. Sci. Comput.*, 21:1657–1669, 2000.

S.C. Brenner. Convergence of the multigrid V-cycle algorithm for second order boundary value problems without full elliptic regularity. *Math. Comp.*, 71:507–525, 2002.

S.C. Brenner. Convergence of nonconforming V-cycle and F-cycle multigrid algorithms for second order elliptic boundary value problem. *Math. Comp.*, 73:1041–1066, 2004.

S.C. Brenner. Lower bounds in domain decomposition. *Proceedings of the 16th International Conference on Domain Decomposition Methods (to appear)*, 2006.

S.C. Brenner and L.R. Scott. *The Mathematical Theory of Finite Element Methods (Second Edition)*. Springer-Verlag, New York-Berlin-Heidelberg, 2002.

S.C. Brenner and L.-Y. Sung. Lower bounds for nonoverlapping domain decomposition preconditioners in two dimensions. *Math. Comp.*, 69:1319–1339, 2000.

F. Brezzi and M. Fortin. *Mixed and Hybrid Finite Element Methods.* Springer-Verlag, New York-Berlin-Heidelberg, 1991.

W.L. Briggs, V.E. Henson, and S.F. McCormick. *A Multigrid Tutorial.* SIAM, Philadelphia, 2001.

C.R. Dohrmann. A preconditioner for substructuring based on constrained energy minimization. *SIAM J. Sci. Comput.*, 25:246–258, 2003.

C. Douglas, G. Hasse, and Langer. *A Tutorial on Elliptic PDE Solvers and Their Parallelization.* Software-Environments-Tools. SIAM, Philadelphia, 2003.

M. Dryja and O.B. Widlund. An additive variant of the Schwarz alternating method in the case of many subregions. Technical Report 339, Department of Computer Science, Courant Institute, 1987.

M. Dryja and O.B. Widlund. Schwarz methods of Neumann-Neumann type for three dimensional elliptic finite element problems. *Comm. Pure Appl. Math.*, 48:121–155, 1995.

H. Elman, D. Silvester, and A. Wathen. *Finite Elements and Fast Iterative Solvers.* Oxford University Press, Oxford, 2005.

C. Farhat, M. Lesoinne, and K. Pierson. A scalable dual-primal domain decomposition method. *Numer. Linear Algebra Appl.*, 7:687–714, 2000.

C. Farhat and F.-X. Roux. A method of finite element tearing and interconnecting and its parallel solution algorithm. *Internat. J. Numer. Methods Engrg.*, 32:1205–1227, 1991.

M. Fortin and R. Glowinski. *Augmented Lagrangian Methods: Applications to to the Numerical Solution of Boundary-Value Problems.* North-Holland, Amsterdam, New York, Oxford, 1983.

A. Greenbaum. *Iterative Methods for Solving Linear Systems.* SIAM, Philadelphia, 1997.

W. Hackbusch. *Multi-grid Methods and Applications.* Springer-Verlag, Berlin-Heidelberg-New York-Tokyo, 1985.

J. Mandel. Balancing domain decomposition. *Comm. Numer. Methods Engrg.*, 9:233–241, 1993.

M.F. Murphy, G.H. Golub, and A.J. Wathen. A note on preconditioning for indefinite linear systems. *SIAM J. Sci. Comput*, 21:1969–1972, 2000.

A. Quarteroni and A. Valli. *Domain Decomposition Methods for Partial Differential Equations.* Clarendon Press, Oxford, 1999.

J.W. Ruge and K. Stüben. Algebraic multigrid. In S.F. McCormick, editor, *Multigrid Methods*, pages 73–130. SIAM, Philadelphia, 1987.

Y. Saad. *Iterative Methods for Sparse Linear Systems.* SIAM, Philadelphia, 2003.

K. Stüben. A review of algebraic multigrid. *J. Comput. Appl. Math.*, 128: 281–309, 2001.

A. Toselli and O.B. Widlund. *Domain Decomposition Methods - Algorithms and Theory.* Springer, New York, 2004.

U. Trottenberg, C. Oosterlee, and A. Schüller. *Multigrid.* Academic Press, San Diego, 2001.

P. Wesseling. *An Introduction to Multigrid Methods.* John Wiley & Sons, Chichester, 1992.

J. Xu. Iterative methods by space decomposition and subspace correction. *SIAM Review*, 34:581–613, 1992.

Numerical Simulation of Charge Transport in Semiconductor Devices Using Mixed Finite Elements

Riccardo Sacco
Dipartimento di Matematica "F. Brioschi"
Politecnico di Milano, via Bonardi 9, 20133 Italy
e-mail: riccardo.sacco@polimi.it
home page: http://www1.mate.polimi.it/~ricsac/

Abstract In this lecture, we present the basic mathematical and numerical tools for semiconductor device simulation using the Drift–Diffusion model. The Gummel's decoupled fixed point map is first reviewed, and then the dual–mixed Formulation of the diffusion–advection–reaction (DAR) differential problems resulting from decoupling is considered. The Galerkin finite element discretization of the weak mixed form is discussed, with a proper treatment of the flux mass matrix that leads to a mixed finite volume approximation of the DAR equation. Several numerical examples are included to validate the proposed procedure.

1 Semiconductor Device Modeling and Simulation

Technological advancement of the semiconductor industry follows the so–called "Moore's Law", which states that every couple of years the number of devices of which a single chip is composed doubles and the dimensions of each device are cut into half. Fig. 1 shows that, up to now, Moore's law has been very accurately verified. To keep this fast and continuous scaling process going on, it is mandatory to resort to numerical simulation to predict the electric properties of devices of the next generation well before their construction is actually feasible. Such simulations must, on the one hand, give quantitatively accurate estimates and, on the other hand, be of a reasonable computational cost so that they can be used within the trial–and–error approach usual in process design or even in automatic optimization tools. The physical description of charge transport on which most of the current simulation tools is based is the Drift–Diffusion (DD) model [6, 15, 9, 10, 8, 5]. A

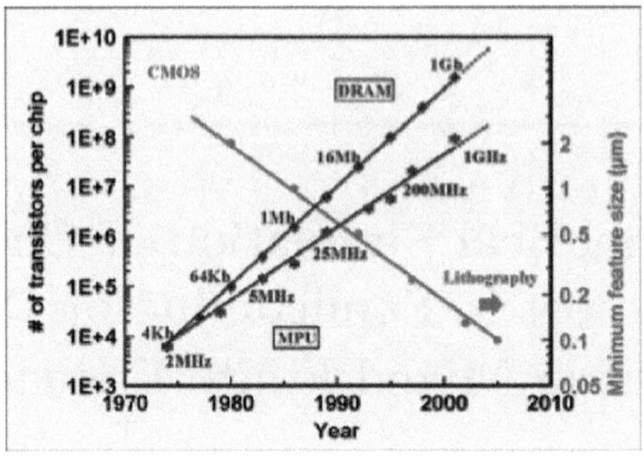

Figure 1. Moore's law in the prediction of advancement of electronic device industry.

striking example is given by Metal Oxided Semiconductor Field Effect Transistors (MOSFETs) which are the basic component of the vast majority of nowadays integrated circuits. Such devices are characterized by a "source" and a "drain" contact and, in between, by a conductive "channel". This latter can be opened or closed by controlling the voltage applied to a third contact called "gate" which is separated from the channel by a thin layer of oxide. Fig. 2 schematically shows the basic structure of a MOSFET and its operation principle.

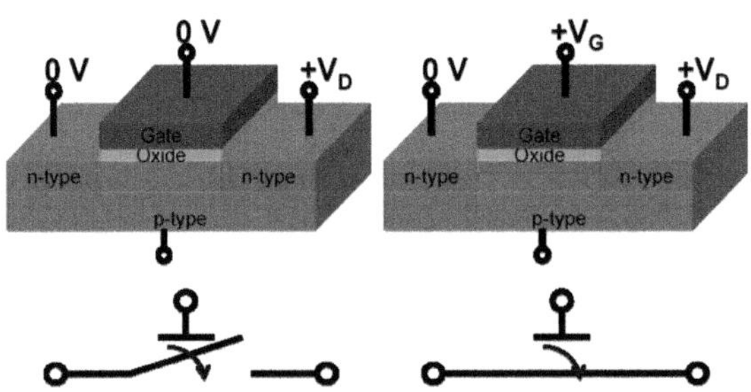

Figure 2. A schematic representation of a a MOSFET device. In the "OFF" state potential barrier prevents electrons from flowing from the "gate" to the "drain" (left). To switch the device into the "ON" state, a voltage must be applied to the "gate", lowering the barrier and allowing current flow (right).

2 The Scaled Drift–Diffusion Model

The scaled and time–dependent Drift–Diffusion (DD) system for describing charge transport in a semiconductor device domain $\Omega \subset \mathbb{R}^d$, with $d \leq 3$, and in the time interval $\mathcal{I}_T = [0, T]$, reads as follows

$$\begin{cases} -\mathrm{div}\,(\lambda^2 \underline{\nabla}\,\varphi) = p - n + N \\ \dfrac{\partial n}{\partial t} - \mathrm{div}\,\underline{J}_n(n, \varphi) = -U(x, \varphi, n, p) \\ \dfrac{\partial p}{\partial t} + \mathrm{div}\,\underline{J}_p(p, \varphi) = -U(x, \varphi, n, p) \\ \underline{J}_n(n, \varphi) = \mu_n\,(\underline{\nabla}\,n - n\underline{\nabla}\,\varphi) \\ \underline{J}_p(p, \varphi) = -\mu_p\,(\underline{\nabla}\,p + p\underline{\nabla}\,\varphi) \\ n = e^{\varphi - \varphi_n}, \qquad p = e^{\varphi_p - \varphi}, \end{cases} \tag{1}$$

where $0 < \lambda << 1$ is the scaled Debye's length. The quantities φ, φ_n and φ_p denote the electric and quasi–Fermi potentials, while n, p, $\underline{J}_n$ and $\underline{J}_p$ are electron and hole concentrations and current densities. The function N is a given datum and represents the doping profile of the device, while U, μ_n and μ_p are the net recombination rate and carrier mobilities. A wide variety of models for these latter quantities can be found in [15], Chapt. 4. Suitable Dirichlet–Neumann boundary conditions and appropriate initial data must be supplied for the DD system (1) to be uniquely solvable (see [15], Chapt. 5).

The immediately evident feature of the DD transport model is represented by its strong nonlinearities and coupling among the equations. Moreover, the presence of the small parameter λ gives rise to singularly perturbed boundary value problems, and the almost discontinuous spatial behaviour of the doping profile N is responsible for markedly heterogeneous solutions (simultaneous presence of steep internal/boundary layers and spatial regions with high smoothness). These considerations are well depicted by Fig. 3, which shows the distribution of potential and carrier concentrations in a simple one–dimensional device, the so-called p-n junction diode.

3 Functional Iterations

In this section, we briefly summarize the classical Gummel's fixed point map that is used for the iterative decoupled solution of system (1) at each time level (see [8] for a thorough analysis).

Let $w^{(0)} = \{\varphi^{(0)}, \varphi_n^{(0)}, \varphi_p^{(0)}\}$ be a given initial datum. Then, the Gum-

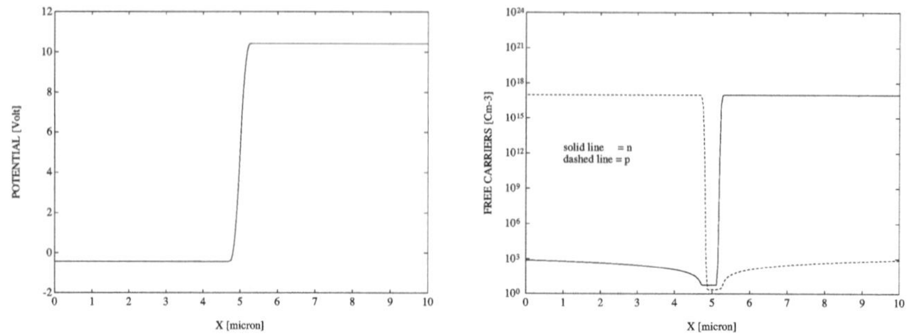

Figure 3. Electric potential φ (left); carrier concentrations n and p (right).

mel's decoupled map consists of solving, for each $k \geq 0$, the following sequence of differential problems:

$$(NLP) \quad \begin{cases} -\mathrm{div}\,(\lambda^2 \,\underline{\nabla}\, \varphi) = e^{\varphi_p^{(k)} - \varphi} - e^{\varphi - \varphi_n^{(k)}} + N & \text{in } \Omega, \\ \varphi = \varphi_D \text{ on } \Gamma_D, \qquad \underline{\nabla}\,\varphi \cdot \underline{n} = 0 \text{ on } \Gamma_N, \end{cases}$$

for $\varphi = \varphi^{(k+1)}$;

$$(LCn) \quad \begin{cases} -\mathrm{div}\, \underline{J}_n(n, \varphi^{(k+1)}) = -U(x, \varphi^{(k+1)}, n, p^{(k)}) & \text{in } \Omega, \\ n = n_D \text{ on } \Gamma_D, \qquad \underline{J}_n(n, \varphi^{(k+1)}) \cdot \underline{n} = 0 \text{ on } \Gamma_N, \end{cases}$$

for $n = n^{(k+1)}$, and a similar equation for $p = p^{(k+1)}$.

Convergence of the overall procedure is monitored by checking that the maximum variations of two successive iterates for φ, φ_n and φ_p becomes sufficiently small.

4 The Diffusion–Advection–Reaction Model Problem

After linearization of (NLP) and (LCn), each differential subproblem of the Gummel iterative procedure can be written in the form of a diffusion–advection–reaction (DAR) boundary value problem

$$\begin{cases} -\mathrm{div}\, \underline{J}(u) + c\, u = f & \text{in } \Omega, \\ \underline{J}(u) = \nu \underline{\nabla}\, u - \underline{b} u & \\ u = g_D & \text{on } \Gamma_D, \\ \underline{J}(u) \cdot \underline{n} = 0 & \text{on } \Gamma_N. \end{cases} \tag{2}$$

In (2), $\nu > 0$ in Ω, g_D is a given boundary datum and $\underline{b}$ is a given advective field; moreover, c and f are positive given functions in Ω in the case of the linearized continuity equations for n and p. Setting $\underline{b} = \underline{0}$ yields the linearized Poisson equation, while setting $\underline{b} = \pm\underline{\nabla}\,\varphi$ yields the linearized continuity equations.

Some important constraints must be satisfied by a sound numerical discretization approach of problem (2):

- a discrete maximum principle (DMP), which ensures:
  - stable approximation of sharp layers (no spurious oscillations) for φ, n and p;
  - positiveness of n and p;
- vector flux conservation (for the electric field $\underline{E} = -\underline{\nabla}\,\varphi$ and the current).

Mixed methods are ideally suited for satisfying the above requests. However, some drawbacks are associated with their use for practical computations:

- mixed methods introduce more unknowns than usual (displacement based) formulations;
- they lead to indefinite saddle–point problems rather than positive definite minimum problems;
- it is not trivial to enforce a DMP when discretizing the DAR problem (2).

5 Dual–Mixed Formulation of the DAR problem

In this section, we introduce an appropriate mixed formulation of the DAR model problem. Recalling that $\underline{b}$ is a gradient field, we introduce the change of variable

$$u = \rho \exp\left(\frac{\underline{b}\cdot\underline{x}}{\nu}\right) := \rho\, e^{\psi}, \tag{3}$$

where ψ is a given potential. Then, system (2) can be written in the equivalent (symmetric) form

$$\begin{cases} -\mathrm{div}\,\underline{J}(\rho) + c\,\rho\, e^{\psi} = f & \text{in } \Omega, \\ \underline{J}(\rho) = \nu\, e^{\psi}\,\underline{\nabla}\,\rho & \\ \rho = g_D\, e^{-\psi} := h_D & \text{on } \Gamma_D, \\ \underline{J}(\rho)\cdot\underline{n} = 0 & \text{on } \Gamma_N. \end{cases} \tag{4}$$

The idea, which is a classical technique dating back to [16, 4], is now to apply the mixed method to the symmetrized form of the DAR problem, and then go back to u using the change of variable (3). With this purpose, let us introduce the following abstract notation

$$Q := L^2(\Omega), \quad V := H_{0,N}(\mathrm{div}\,;\Omega) = \{\underline{\tau} \in H(\mathrm{div}\,;\Omega) \,|\, \underline{\tau}\cdot\underline{n} = 0 \,\mathrm{on}\, \Gamma_N\},$$

$$a(\underline{J},\underline{\tau}) = \int_\Omega (\nu\, e^{\psi})^{-1}\, \underline{J}\cdot\underline{\tau}\, dx,$$

$$b(q,\underline{\tau}) = \int_\Omega q\, \mathrm{div}\, \underline{\tau}\, dx, \quad c(v,q) = \int_\Omega c\, e^{\psi}\, v\, q\, dx$$

$$F(\underline{\tau}) = \int_{\Gamma_D} h_D\, \underline{\tau}\cdot\underline{n}\, d\gamma, \quad G(q) = -\int_\Omega f\, q\, dx.$$

Then, the dual–mixed (DM) formulation of (4) reads:
Find $\rho \in Q$ and $\underline{J} \in V$ such that

$$\begin{cases} a(\underline{J},\underline{\tau}) + b(\rho,\underline{\tau}) &= F(\underline{\tau}) \qquad \forall \underline{\tau} \in V \\ b(q,\underline{J}) - c(\rho,q) &= G(q) \qquad \forall q \in Q. \end{cases} \tag{5}$$

It can be proved that problem (5) admits a unique solution (see [3]).

6 Galerkin Finite Element Approximation

Upon introducing two suitable finite dimensional spaces

$$Q_h \subset Q, \qquad V_h \subset V,$$

the Galerkin finite element approximation of (5) reads:
Find $\rho_h \in Q_h$ and $\underline{J}_h \in V_h$ such that

$$\begin{cases} a(\underline{J}_h,\underline{\tau}_h) + b(\rho_h,\underline{\tau}_h) &= F(\underline{\tau}_h) \qquad \forall \underline{\tau}_h \in V_h \\ b(q_h,\underline{J}_h) - c(\rho_h,q_h) &= G(q_h) \qquad \forall q_h \in Q_h. \end{cases} \tag{6}$$

In matrix form, the above problem is

$$\begin{cases} A\,\mathbf{J} + B^T\boldsymbol{\rho} &= \mathbf{f} \\ B\,\mathbf{J} - C\,\boldsymbol{\rho} &= \mathbf{g}. \end{cases}$$

6.1 Approximation Using Raviart–Thomas Elements

Let $\mathcal{T}_h$ be a regular triangulation of Ω into elements K [7]. The usual (and computationally less expensive!) choice is to employ the Raviart–Thomas finite element space of lowest degree [12]

$$Q_h(K) = \mathbb{P}_0(K), \qquad V_h(K) = \mathbb{RT}_0(K) = (\mathbb{P}_0)^2 \oplus \underline{x}\,\mathbb{P}_0(K) \qquad \forall K \in \mathcal{T}_h,$$

in such a way that the global approximation spaces are defined as

$$Q_h = \{q \in Q \,|\, q|_K \in Q_h(K) \; \forall K \in \mathcal{T}_h\},$$

$$V_h = \{\underline{\tau} \in V \,|\, \underline{\tau}|_K \in V_h(K) \; \forall K \in \mathcal{T}_h\}.$$

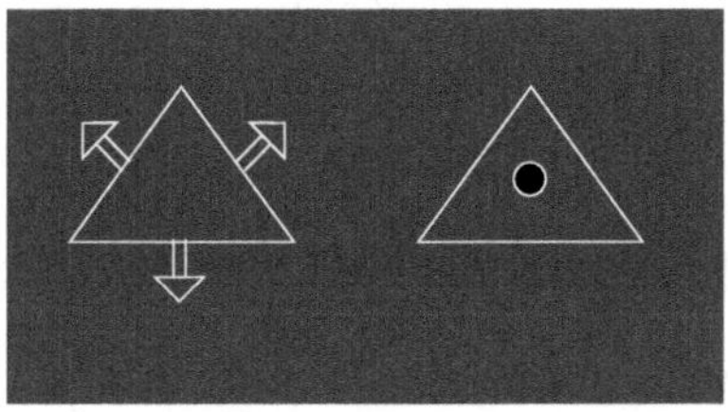

Figure 4. Degrees of freedom of the RT finite element space of lowest order.

The basic properties of the discrete DM Formulation can be summarized as follows:

- the discrete current $\underline{J}_h$ has continuous normal components across interelement edges;
- optimal error estimates can be proved for ρ_h and $\underline{J}_h$;
- matrix A is square and positive definite;
- however, A^{-1} is full and the system matrix of (6) is not, in general, positive definite, unlike the case of the standard Galerkin primal–based discretization (refer to Prof. Brenner's lectures on fast solution of the saddle–point problem (6));
- no DMP irrespective, in general, of the value of c can be proved for ρ_h.

To cure the drawbacks of the DM formulation, several alternative approaches can be pursued:

(A) relax the constraint $V_h \subset V$ and enforce it back through the introduction of a Lagrange multiplier. This procedure is called hybridization, and for further details we refer to [1] and the Lecture about Process Device Simulation;

(B) eliminate **J** from the first equation of (6) through a (suitable) diagonalization of matrix A. This technique has been proposed and analyzed in [11, 5] and leads to a variant of the DM method, called Mixed Finite Volume (MFV) formulation.

6.2 Elimination of the Current J

Using (B), we approximate A with a diagonal matrix A_h, in such a way that

$$\mathbf{J} = A_h^{-1}(\mathbf{f} - B^T \boldsymbol{\rho}).$$

Consequently, the final system acting on the primal variable $\boldsymbol{\rho}$ reads

$$M_\rho \boldsymbol{\rho} = \mathbf{b}_\rho, \tag{7}$$

with

$$M_\rho = (B\, A_h^{-1}\, B^T + C), \quad \mathbf{b}_\rho = B\, A_h^{-1}\mathbf{f} - \mathbf{g}.$$

We are going to show that a proper choice of the diagonal matrix A_h allows to:

- obtain a symmetric and positive definite matrix M_ρ, such that $(M_\rho^{-1})_{ij} > 0$ (M-matrix property);
- ensure a DMP for ρ_h irrespective of c;
- maintain optimal convergence rates of the DM formulation without diagonalization of the mass matrix.

For each edge $\underline{e}_i \notin \Gamma$, we denote by:

- T^+ and T^- the two triangles of $\mathcal{T}_h$ which share $\underline{e}_i$;
- C^+ and C^- the corresponding circumcenters;
- $d_i = dist(C^+, C^-)$;
- ρ^+ and ρ^- the values of ρ_h over T^+ and T^-;
- $\mathcal{J}_i$ the flux of $\underline{J}_h$ through $\underline{e}_i$;
- $\underline{\tau}_i$ the RT test function associated with $\underline{e}_i$.

If $\underline{e}_i \in \Gamma$, then $d_i = dist(C^+, \Gamma)$. Then, for each edge $\underline{e}_i \notin \Gamma_N$, we set

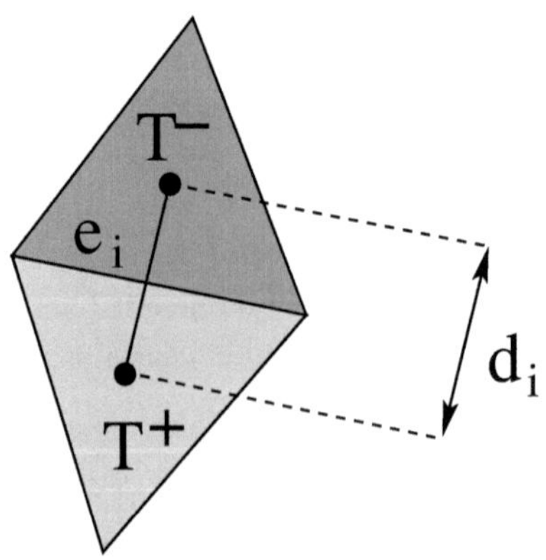

Figure 5. Geometrical notation for the MFV formulation.

$$a(\underline{J}_h, \underline{\tau}_i) \Rightarrow a_h(\underline{J}_h, \underline{\tau}_i) = \overline{\nu\, e^{\psi}}_i^{-1} \mathcal{J}_i \frac{d_i}{|\underline{e}_i|},$$

while for each edge $\underline{e}_i \in \Gamma_N$ we directly set

$$\mathcal{J}_i = 0.$$

The quantity $\overline{\nu\, e^{\psi}}_i$ is the harmonic average of the function $\nu\, e^{\psi}$ along the interval connecting C^+ and C^-, defined as

$$\overline{\nu\, e^{\psi}}_i = \frac{d_i}{\displaystyle\int_{C^+}^{C^-} (\nu\, e^{\psi})^{-1}\, ds}.$$

It is well-known that the use of harmonic averaging in the discretization of elliptic problems with discontinuous (and strongly varying) coefficients provides superior accuracy and stability compared with standard averages (see [2]).

Using the previous approximation, and observing that $\operatorname{div} \underline{\tau}_i = 1/|T^+|$ on T^+ and $\operatorname{div} \underline{\tau}_i = -1/|T^-|$ on T^-, we eventually get the following explicit expression for the current flux across edge $\underline{e}_i$

$$\mathcal{J}_i = \begin{cases} \overline{\nu\, e^{\psi}}_i \dfrac{\rho^- - \rho^+}{d_i} |\underline{e}_i| & \forall \underline{e}_i \notin \Gamma_N \\ 0 & \forall \underline{e}_i \in \Gamma_N. \end{cases} \tag{8}$$

7 Mixed Finite Volume Formulation

Substituting (8) into the second equation of (6), we obtain the following Mixed Finite Volume (MFV) equation for ρ_h to be satisfied $\forall K_k \in \mathcal{T}_h$

$$\sum_{\substack{r \in E(k) \\ e^r \notin \Gamma_N}} \overline{\nu\, e^{\psi}}_r \left(\frac{\rho_k^+ - \rho_{j(r)}^-}{d_r} \right) |\underline{e}_r| + c_k\, e^{\psi_k} \rho_k^+ |K_k| = f_k |K_k|, \tag{9}$$

where, for an edge $\underline{e}_r \notin \Gamma_N$, $j(r)$ is the index of the triangle T^- sharing $\underline{e}_r$ and $\rho_{j(r)}^-$ is the unknown associated with T^-. Notice that if $\underline{e}_r \in \Gamma_D$, we set directly $\rho_{j(r)}^- = h_{D,r}$ (equal to the average of h_D on $\underline{e}_r$), while we set directly $\mathcal{J}_r = 0$ if $\underline{e}_r \in \Gamma_N$. The MFV scheme associated with (9) is of the form (7), and its can be readily checked that:

- matrix M_ρ is symmetric positive definite, with strictly positive inverse (Stiltjes matrix) [17]. This immediately implies that a DMP holds for ρ_h;

- the stencil of the MFV method (9) is compact (at most, we have 4 nonzero entries for each row of M_ρ!);
- the same optimal error estimates valid for the DM method still hold for the MFV variant, i.e.

$$\|\rho - \rho_h\|_{L^2(\Omega} + \|\underline{J} - \underline{J}_h\|_{H(\mathrm{div}\,;\Omega)} \leq \mathcal{C}\, h, \tag{10}$$

where $\underline{J} \equiv \underline{J}(\rho)$ and $\mathcal{C}$ is a positive constant depending on suitable norms of the solution pair $(\rho,\, \underline{J})$ of (4) and independent of the mesh size h;
- current flux conservation across interelement edges is still enforced.

Computational difficulties due to the presence of exponential terms in (9) (machine overflow/underflow) suggest to go back to the original variable u using the transformation

$$\rho_k = u_k\, e^{-\psi_k} \qquad \forall K_k \in \mathcal{T}_h.$$

Using this (discrete) change of variable leads to solving for the unknown u_h the following MFV system $\forall K_k \in \mathcal{T}_h$

$$\sum_{\substack{r \in E(k)\\ e^r \notin \Gamma_N}} \overline{\nu}_r \left(\frac{B(\Delta\psi_{j(r),k}) u_k^+ - B(-\Delta\psi_{j(r),k}) u_{j(r)}^-}{d_r} \right) |\underline{e}_r| + c_k\, u_k\, |K_k| = f_k\, |K_k|, \tag{11}$$

where $\Delta\psi_{j(r),k} = \psi_{j(r)} - \psi_k$ and

$$B(x) = \begin{cases} \dfrac{x}{e^x - 1} & \text{if } x \neq 0 \\ 1 & \text{if } x = 0. \end{cases}$$

System (11) can be written in matrix form as

$$M_u \mathbf{u} = \mathbf{b}_u$$

where:

- matrix M_u is no longer symmetric positive definite as in (9);
- however, it can be checked that M_u is strictly diagonally dominant by columns;
- moreover, since $(M_u)_{ij} \leq 0$, $i \neq j$ and $(M_u)_{ii} > 0$, $i = j$, it turns out that M_u is an M–matrix (M_u^{-1} is a positive matrix);
- this implies that a DMP is satisfied by the nonsymmetric MFV method;
- finally, the MFV scheme is an exponentially fitted difference method, and can be regarded as a 2D extension of the classical Scharfetter–Gummel approximation, valid in the one–dimensional case (see [14, 13]).

8 Numerical Results

In this section, we provide an extensive validation of the MFV scheme (9) through the numerical study of several DAR problems exhibiting solutions with sharp internal and boundary layers. In all of the three considered test cases, the computational domain is $\Omega = (0,1)^2$, while the last test case deals with the simulation of a real–life MOS structure.

8.1 Test Case 1

This test case describes a DAR model problem with a strongly varying diffusion coefficient, but with a smooth solution. With this purpose, we set $\underline{b} = \underline{0}$, $c = 1$ and homogeneous Dirichlet b.c.s for u in (2). Moreover, we set

$$\nu(x,y) = \left(\frac{\nu_a + \nu_b}{2}\right)\left(1 + \frac{\nu_b - \nu_a}{\nu_a + \nu_b}\tanh(K(y-x))\right),$$

where $\nu_b > \nu_a > 0$ and K is a suitable constant. For $\nu_b/\nu_a \gg 1$ and $K \gg 1$, ν tends to a step function with jump equal to $\nu_b - \nu_a$ across $y = x$. The right–hand side f is chosen in such a way that the exact solution is

$$u(x,y) = 16x\,y\,(x-1)\,(y-1).$$

Figs. 6–9 show the obtained results corresponding to two different sets of values of the problem parameters. The computed errors are in good agreement with theoretical estimates and exhibit optimal convergence rates as functions of h.

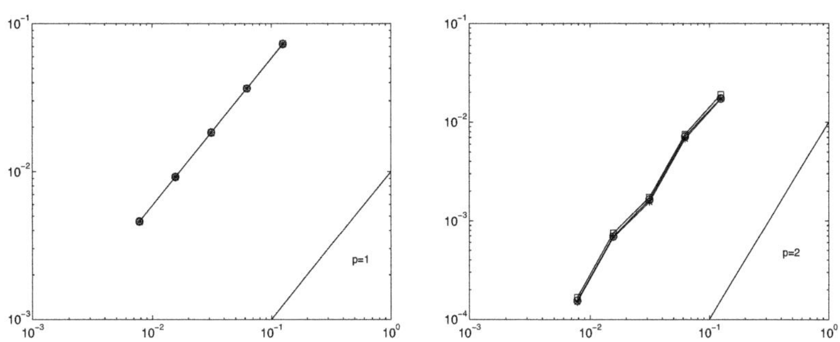

Figure 6. $||u - u_h||_{L^2(\Omega)}$ (left) and $||u - u_h||_{L^\infty(\Omega)}$ (right) ($\nu_a = 1$, $\nu_b = 10$, $K = 2$).

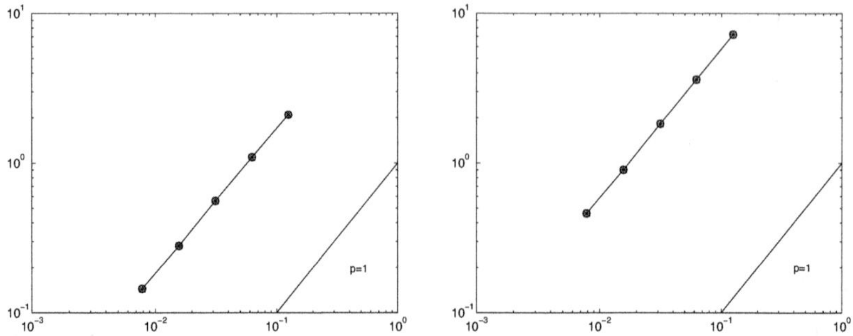

Figure 7. $||\underline{J} - \underline{J}_h||_{L^2(\Omega)}$ (left) and $||U - U_h||_{H(\text{div};\Omega)\times L^2(\Omega)}$ (right), where $U - U_h = (u - u_h, \underline{J} - \underline{J}_h)$ ($\nu_a = 1$, $\nu_b = 10$, $K = 2$).

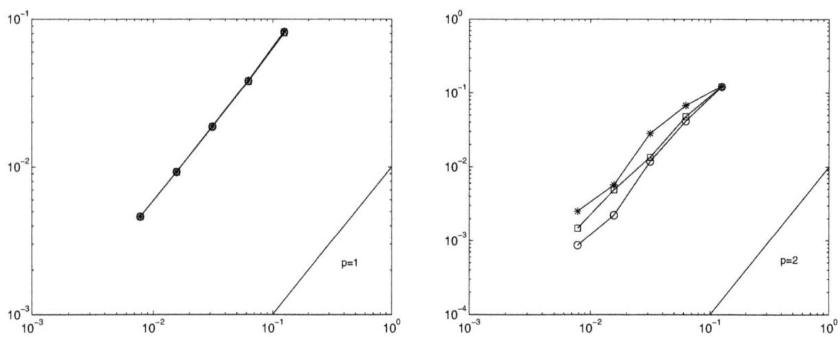

Figure 8. $||u - u_h||_{L^2(\Omega)}$ (left) and $||u - u_h||_{L^\infty(\Omega)}$ (right) ($\nu_a = 1$, $\nu_b = 1000$, $K = 100$).

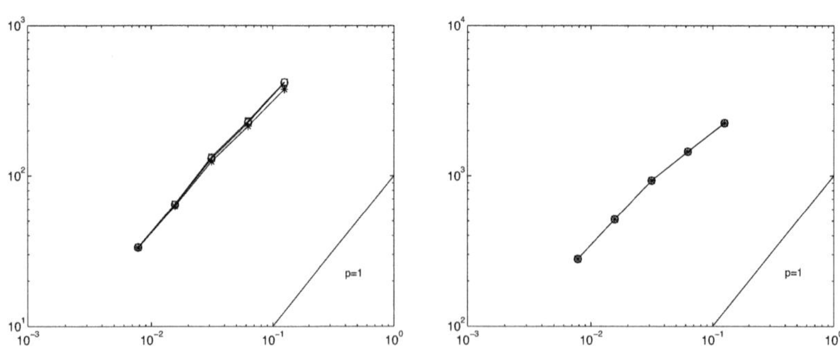

Figure 9. $||\underline{J} - \underline{J}_h||_{L^2(\Omega)}$ (left) and $||U - U_h||_{H(\text{div};\Omega)\times L^2(\Omega)}$ (right) ($\nu_a = 1$, $\nu_b = 1000$, $K = 100$).

8.2 Test Case 2

This test case describes a DAR model problem in the reaction–dominated regime, and aims at emphasizing the ability of the MFV formulation in handling this situation without introducing numerical instabilities, unlike what happens with the standard DM method using the Raviart–Thomas space of lowest order. We set $\underline{\nabla}\,\psi = (1, 0.5)^T$, $f = c = 10^6$, homogeneous Dirichlet boundary conditions for u and diffusion coefficient $\nu = 1$. Figure 10 shows the solution computed by the two above mentioned approaches.

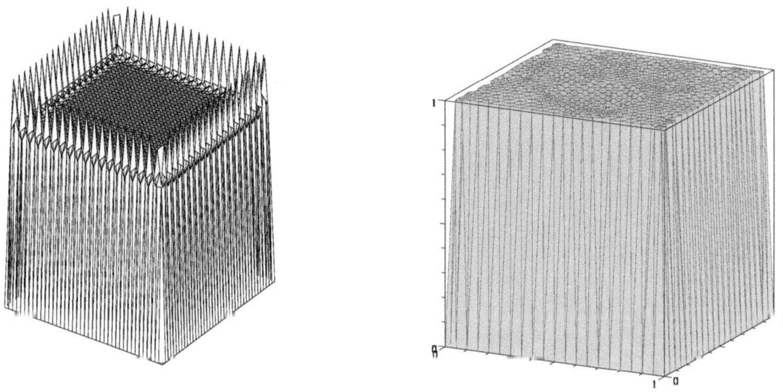

Figure 10. Computed u_h: standard DM solution using $\mathbb{RT}_0$ finite elements (left), MFV solution (right).

8.3 Test Case 3

This test case describes a DAR model problem in a semiconductor–like format, and aims at emphasizing the ability of the MFV formulation in handling sharp internal layers and in correctly approximating current flow in the computational domain. The Dirichlet boundary conditions are set as follows

$$u = 10^3 \quad \text{for } 0 \le x \le 0.2,\ y = 0 \text{ and } x = 0\,,\ 0 \le y \le 0.2;$$

$$u = 10^{17} \quad \text{for } x = 1\,,\ 0.8 \le y \le 1 \text{ and } 0.8 \le x \le 1\,,\ y = 1.$$

On Γ_N homogeneous Neumann conditions are given, while $c = 0$ and $\psi(x, y) = \psi_0(x, y)/\ell$, with

$$\psi_0(r) = \begin{cases} 0 & 0 \le r \le 0.8, \\ 2(r - 0.8) & 0.8 \le r \le 0.9, \\ 0.2 & r \ge 0.9, \end{cases} \tag{12}$$

where $r = (x^2 + y^2)^{1/2}$, and $\ell = 10^{-4}$. The right–hand side f is equal to 0.5×10^{16} in the region $0.82 \leq \rho \leq 0.88$, while $f = 0$ elsewhere. These data are chosen in such a way to represent a curved $p-n$ junction diode subject to a high reverse bias. Figure 11 shows the computed solutions u_h and $\underline{J}_h$. Results demonstrate that the steep layers for u_h are well captured without spurious oscillations and that the current distribution exhibits the typical "crowding" effect around the Dirichlet boundary at the upper right corner of the domain.

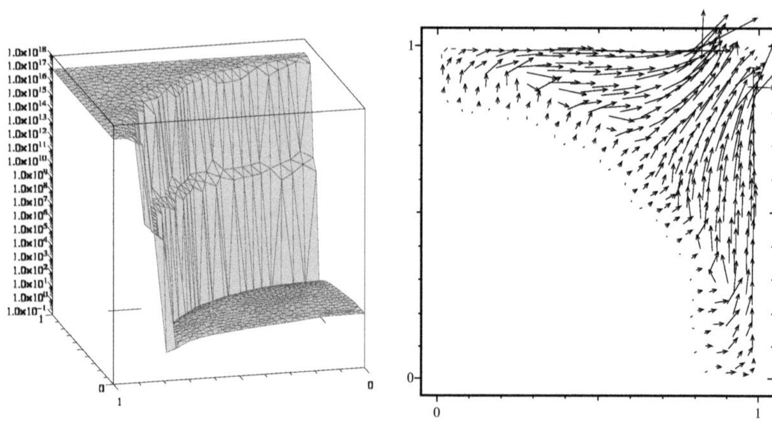

Figure 11. Computed u_h, log–scale (left), computed $\underline{J}_h$ (right).

8.4 Test Case 4

This last test case deals with the simulation of a real–life semiconductor device. A modern submicron MOS transistor is considered in Fig. 12, which shows the geometry of the problem and the doping profile N. Figs. 13–14 show the computed solutions. The results clearly demonstrate that the steep layers in the electron and hole densities are well captured without introducing spurious oscillations. Moreover, a very high peak of the electric field is present at the drain end of the channel. This electric field is responsible for the high acceleration experimented by the electrons throughout the channel region. Finally, a strong carrier thermal generation arises at the entrance of the channel, due to the reverse–biasing conditions between the source and bulk regions of the device.

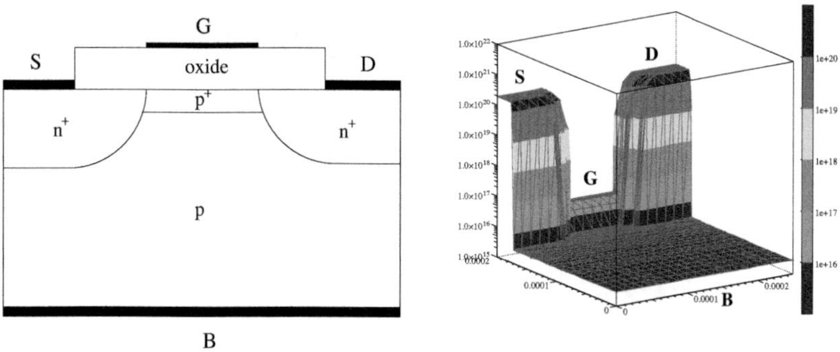

Figure 12. A MOSFET device: geometry (left), doping profile, log–scale (right).

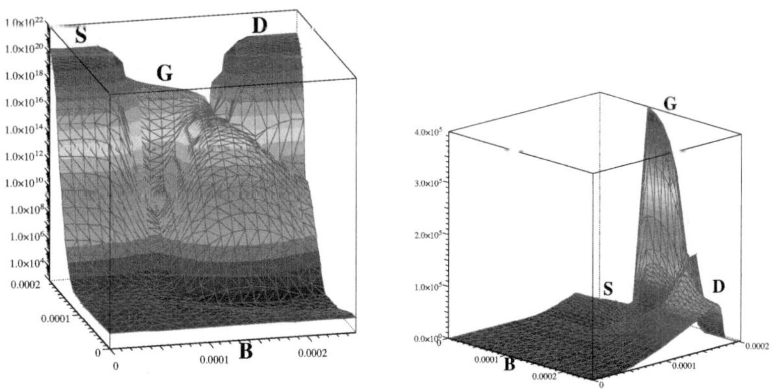

Figure 13. Electron density, log–scale (left) and electric field strength (right).

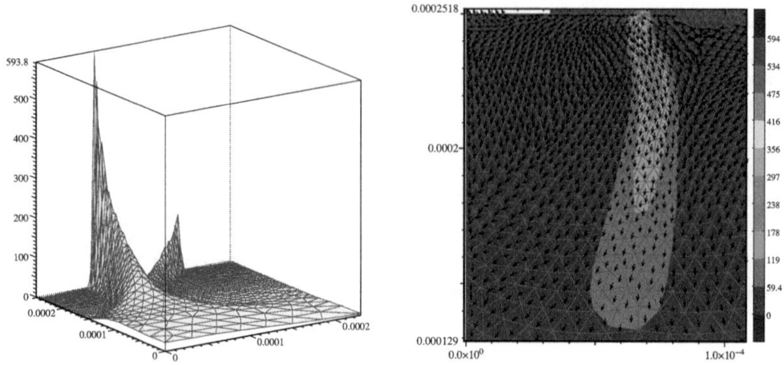

Figure 14. Hole current density: module (left) and vector plot (right).

Bibliography

[1] D. N. Arnold, F. Brezzi, Mixed and nonconforming finite element methods: implementation, postprocessing and error estimates, RAIRO Modél. Math. Anal. Numér., 19, 7–32 (1995).

[2] I. Babuška, J.E. Osborn, Generalized Finite Element Methods: Their Performance and Their Relation to Mixed Methods, SIAM J. Numer. Anal., 20 (3), 510–536 (1983).

[3] F. Brezzi, M. Fortin, *Mixed and Hybrid Finite Element Methods*, Springer-Verlag, New York (1991).

[4] F. Brezzi, P. Marini, P. Pietra, Two-dimensional exponential fitting and applications to semiconductor device equations, SIAM J. Numer. Anal. 26, 1342–1355 (1989).

[5] F. Brezzi, L. D. Marini, S. Micheletti, P. Pietra, R. Sacco, S. Wang, Discretization of Semiconductor Device Problems (I), 317–441, Handbook of Numerical Analysis, Vol. XIII, P.G. Ciarlet Ed., W.H.A. Schilders, E.J.W. ter Maten, Guest Eds., Elsevier North-Holland, Amsterdam (2005).

[6] E. Buturla, P. Cottrell, B.M. Grossman, K.A. Salsburg, Finite Element Analysis of Semiconductor Devices: The FIELDAY Program, IBM J. Res. Develop., 25 (4), 218–231 (1981).

[7] P.G. Ciarlet *The finite element method for elliptic problems*, North-Holland, Amsterdam (1978).

[8] J.W. Jerome, *Analysis of Charge Transport*, Springer-Verlag Berlin Heidelberg (1996).

[9] P. A. Markowich, *The Stationary Semiconductor Device Equations*, Springer-Verlag, Wien-New York (1986).

[10] P. A. Markowich, C. A. Ringhofer, C. Schmeiser, *Semiconductor equations*, Springer-Verlag, Wien New York, 1990.

[11] S. Micheletti, R. Sacco, F. Saleri, On Some Mixed Finite Element Methods with Numerical Integration, SIAM J. Sci. Comput. 23, No.1, 245-270 (2001).

[12] P.A. Raviart, J.M. Thomas, A mixed finite element method for second order elliptic problems, in Mathematical Aspects of the Finite Element Method (I. Galligani, E. Magenes, eds.), Lecture Notes in Math., Springer-Verlag, New York, 606, 292–315 (1977).

[13] H. G. Roos, M. Stynes, L. Tobiska, Numerical methods for singularly perturbed differential equations, Springer-Verlag, Berlin Heidelberg (1996).

[14] D.L. Scharfetter, H.K. Gummel, Large signal analysis of a silicon Read diode oscillator. IEEE Trans. Electron Devices ED-16, 64–77 (1969).

[15] S. Selberherr, *Analysis and Simulation of Semiconductor Devices*, Springer-Verlag, Wien-New York (1984).

[16] J. W. Slotboom, Computer-Aided Two-Dimensional Analysis of Bipolar Transistors, IEEE Trans. Electr. Dev., ED-20, 669–679 (1973).

[17] R.S. Varga, *Matrix Iterative Analysis*, Prentice–Hall, Englewood Cliffs, New Jersey (1962).

Numerical Simulation of Thermal Oxidation Process in Semiconductor Devices Using Mixed–Hybrid Finite Elements

Riccardo Sacco

Dipartimento di Matematica "F. Brioschi"
Politecnico di Milano, via Bonardi 9, 20133 Italy
e-mail: riccardo.sacco@polimi.it
home page: http://www1.mate.polimi.it/~ricsac/

Abstract In this lecture, we present the basic mathematical and numerical tools for the simulation of the thermal oxidation process in semiconductor device technology. The mathematical model is first reviewed, emphasizing that it requires at each time level the solution of a diffusion-reaction problem and a fluid-structure interaction problem. Then, mixed-hybrid finite elements are introduced for the numerical approximation of each differential subproblem, with a detailed discussion of the static condensation procedure to eliminate the mixed variables in favor of the hybrid Lagrange multipliers. Several numerical examples are included to validate the accuracy and stability of the proposed computational procedure.

1 The Thermal Oxidation Process

The up-to-date technology in silicon integrated circuits is characterized by strongly heterogeneous mechanical properties, which develop during fabrication stresses that can lead to current leakages and short-circuits. One of the main sources of these negative features is the thermal oxidation process, that typically occurs in the isolation of adjacent electrically active zones on a semiconductor chip and in the fabrication of Metal Oxide Semiconductors devices (MOS). Thermal oxidation comprises a highly complex sequence of chemical and fluid–mechanical intermediate steps, so that numerical simulation can provide an important tool to optimize the design strategy [10, 14].

The overall oxidation procedure can be modeled as a fluid-structure interaction problem driven by the oxygen diffusion and reaction process [15, 16, 6]. In the oxidation process, a layer of silicon dioxide (SiO_2) is grown at high temperature on a silicon wafer bulk (Si) partially masked by a silicon-nitride (Si_3N_4) pattern impermeable to the oxidant penetration. The oxygen diffuses through the oxide and reacts with the silicon at the Si-SiO_2 interface (see Fig. 1). The SiO_2 has a molar volume 2.2 times larger

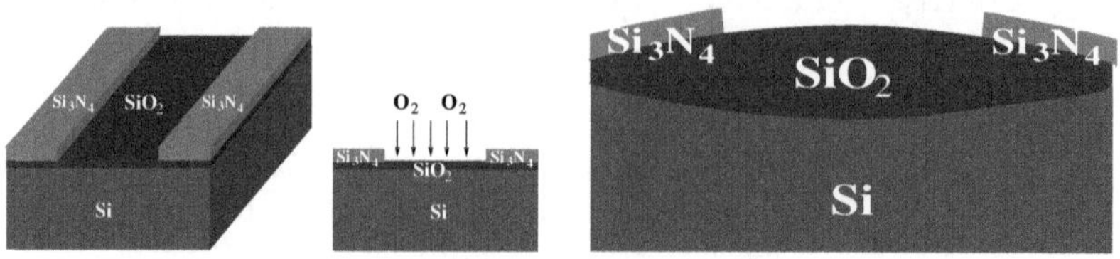

Figure 1. The LOCal Oxidation Structure (LOCOS).

than Si, which means that at each time level a volume fraction of the new grown oxide replaces the silicon that has been consumed, while the remaining volume fraction pushes the old oxide upward. This constrained volume expansion gives rise to large stresses and causes the SiO_2 to undergo a compressive state and the Si to undergo a tensile state. Experimental evidence suggests that this state of stress significantly affects both the O_2 diffusion in the pre-existing SiO_2 layer and also the chemical reaction kinetics between the O_2 and the Si. The combination of these effects alters the final shape of the Si-SiO_2 system that presents non planar regions (even when the oxide grows on planar regions) and gives rise to the so-called "bird's beak" region (see Fig. 2).

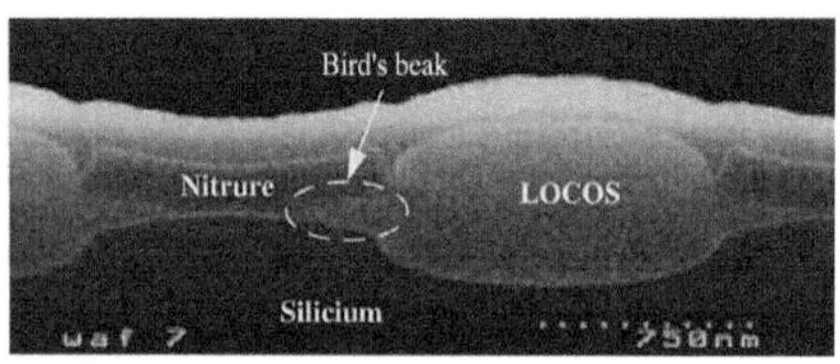

Figure 2. Bird's beak oxide region.

2 An Overview of the Coupled Problem

The mathematical model of the process is constituted by a sequence of quasi-stationary steps each involving the solution of two PDE systems over a domain whose shape is changing in time. The first system is a diffusion-reaction problem for the oxidant and the second system is a stress analysis problem in the oxide, nitride and silicon bulk. The two PDE systems are mutually dependent because:

- the diffusion and kinetic reaction coefficients as well as the geometry of the deformed domain depend on the stress distribution;
- in turn, the chemical reaction forces the oxide-silicon interface to move, driving the mechanical problem.

A complex coupled problem is therefore obtained, consisting of a fluid-structure interaction problem driven by a diffusion-reaction process (see the schematic diagram of Fig. 3). It is important to notice that the data

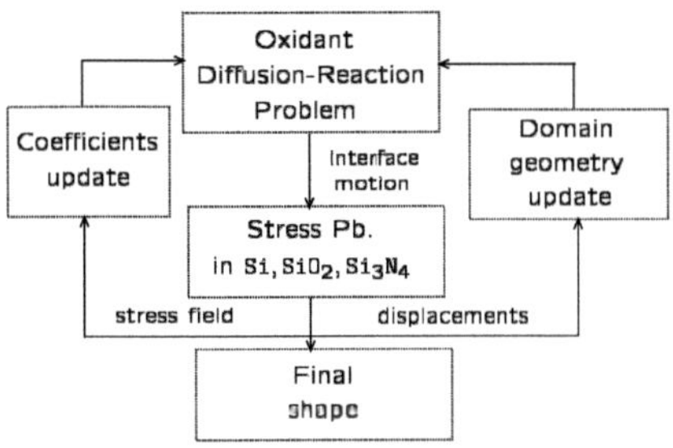

Figure 3. Block scheme of the process simulation algorithm.

exchanged between the sub-blocks in both the coupled systems are fluxes and stress dependent quantities, so that the quality of the approximation of these latter quantities clearly affects the accuracy of the overall computation. In particular, a sound numerical approach must account for the following issues:

- the two subproblems involved in thermal oxidation are mutually dependent through the exchange of stresses and fluxes at material interfaces;
- the materials involved in the fluid–structural problem are strongly heterogeneous (Si and Si_3N_4: compressible elastic, SiO_2: incompressible visco–elastic)

Based on the above constraints, the use of mixed-hybrid finite element formulations seems to be appropriate because:

- stresses and fluxes are directly computed quantities, rather than obtained from post-processing techniques;
- the incompressible regime can be authomatically recovered within a unified formulation (no need of using two separate computer codes!!).

3 Hybrid Formulations

In this section, we provide a short and self–consistent presentation of hybrid formulations. For further information and details of the proofs, see [17, 2, 19, 3].

Given $f \in L^2(\Omega)$, let us consider the following minimization problem on $H_0^1(\Omega)$

$$\inf_{v \in H_0^1(\Omega)} \mathcal{J}(v), \qquad \mathcal{J}(v) = \frac{1}{2}\int_\Omega |\nabla v|^2 \, dx - \int_\Omega f\, v\, dx.$$

The functional $\mathcal{J}(v)$ is the potential energy of the system and its unique minimizer is a function $u \in H_0^1(\Omega)$ such that

$$\int_\Omega \nabla u \cdot \nabla v \, dx = \int_\Omega f\, v\, dx, \qquad \forall v \in H_0^1(\Omega).$$

The solution u satisfies in the sense of distributions the Poisson equation

$$\begin{cases} -\Delta u = f & in\ \Omega, \\ u = 0 & on\ \Gamma. \end{cases} \tag{1}$$

The above problem is the primal weak formulation and u is the primal unknown. Let us now consider a general partition $\mathcal{T}_h$ of $\overline{\Omega}$ into subdomains

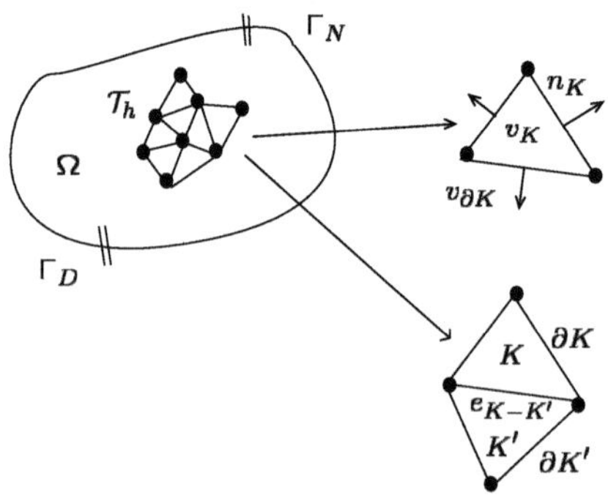

Figure 4. Geometrical discretization and notation.

K (for example, triangles) such that

$$\overline{\Omega} = \bigcup_{K \in \mathcal{T}_h} \overline{K}.$$

Hybrid formulations are characterized by some common points:

- the variational principle from which they emanate depends explicitly on a partition of the domain $\overline{\Omega}$ into subdomains;
- variables defined in the interior of each subdomain coexist with variables defined on the boundary of each subdomain (hybrid variables).

Using the partition $\mathcal{T}_h$, the energy functional $\mathcal{J}(v)$ can be written as

$$\mathcal{J}(v) = \frac{1}{2}\int_\Omega |\nabla v|^2\,dx - \int_\Omega f\,v\,dx = \sum_{K\in\mathcal{T}_h}\left\{\frac{1}{2}\int_K |\nabla v|^2\,dx - \int_K f\,v\,dx\right\}.$$

Introducing the *broken* space

$$W = \{v\,|\,v \in L^2(\Omega),\, v_K \in H^1(K)\ \forall K \in \mathcal{T}_h\} = \prod_{K\in\mathcal{T}_h} H^1(K),$$

we can extend the functional $\mathcal{J}(v)$ on W. The space $H_0^1(\Omega)$ is a closed subspace of W that can be characterized as the set of functions $v \in W$ such that

$$\sum_{K\in\mathcal{T}_h}\int_{\partial K} v\,q\cdot n\,ds = 0 \qquad \forall q \in H(\mathrm{div}\,;\Omega).$$

This relation expresses the continuity of a function in $H_0^1(\Omega)$ at interfaces $e_{K-K'}$ $\forall K, K' \in \mathcal{T}_h$, and the condition $v_\Gamma = 0$. The variable $q\cdot n$ is the Lagrange multiplier (hybrid variable) that enforces the continuity constraint for v at interelement boundaries.

3.1 Primal-Hybrid Formulation

Let $\widetilde{Q} = H(\mathrm{div}\,;\Omega)$. We consider the saddle-point problem

$$\inf_{v\in W}\ \sup_{q\in\widetilde{Q}}\ \mathcal{L}_{PH}(v,q) = \sum_{K\in\mathcal{T}_h}\left(\frac{1}{2}\int_K |\nabla v|^2\,dx - \int_{\partial K} v\,q\cdot n\,ds\right) - \int_\Omega f\,v\,dx,$$

for which the optimality conditions read:
find $(u,p) \in (W \times \widetilde{Q})$ such that

$$\begin{cases} \displaystyle\sum_{K\in\mathcal{T}_h}\left(\int_K \nabla u\cdot\nabla v\,dx - \int_{\partial K} v\,p\cdot n\right) = \int_\Omega f\,v\,dx & \forall v \in W, \\ \displaystyle\sum_{K\in\mathcal{T}_h}\int_{\partial K} u\,q\cdot n\,ds = 0 & \forall q \in \widetilde{Q}. \end{cases}$$

The above problem is the primal hybrid (PH) formulation of the elliptic model problem (1). It is clear that only the normal traces of $p \in \widetilde{Q}$ can be

determined. For this reason, we can equivalently reformulate the primal-hybrid problem as:
find $(u, p \cdot n) \in (W \times Q)$ such that

$$\begin{cases} \displaystyle\sum_{K \in \mathcal{T}_h} \left(\int_K \nabla u \cdot \nabla v \, dx - \int_{\partial K} v \, p \cdot n \right) = \int_\Omega f \, v \, dx & \forall v \in W, \\ \displaystyle\sum_{K \in \mathcal{T}_h} \int_{\partial K} u \, q \cdot n \, ds = 0 & \forall q \cdot n \in Q, \end{cases}$$

where

$$Q = \left\{ \mu \in \prod_{K \in \mathcal{T}_h} H^{-1/2}(\partial K) \,|\, \exists \tau \in H(\mathrm{div}\,;\Omega) \,|\, \tau \cdot n = \mu \,\mathrm{on}\, \partial K, \forall K \in \mathcal{T}_h \right\}.$$

3.2 Approximation Properties of PH Method

Let us introduce the finite dimensional spaces

$$W_h = \left\{ v_h \in W \,|\, v_h^K \in \mathbb{P}_1(K) \, \forall K \in \mathcal{T}_h \right\},$$
$$Q_h = \left\{ q_h \in Q \,|\, q_h^K \in \mathbb{RT}_0(K) \, \forall K \in \mathcal{T}_h \right\}.$$

Using the abstract error analysis of Babuska–Brezzi theory and standard interpolation results, we have the optimal error estimate

$$|u - u_h|_{1,\Omega} + ||p \cdot n - p_h \cdot n||_{-1/2,\Omega} \le Ch|u|_{2,\Omega}.$$

The primal-hybrid approximate solution u_h is actually continuous at the midpoint of each single edge of $\mathcal{T}_h$. Therefore, it is numerically convenient to deal with the primal-hybrid formulation by solving the following nonconforming approximation of (1):
find $u_h^{NC} \in W_{h,0}^{NC}$ such that we have

$$\sum_{K \in \mathcal{T}_h} \int_K \nabla u_h^{NC} \cdot \nabla v_h \, dx = \sum_{K \in \mathcal{T}_h} \int_K f \, v_h \, dx, \qquad \forall v_h \in W_{h,0}^{NC},$$

where for any function $g \in H^{1/2}(\Gamma)$ we define $W_{h,g}^{NC}$ as the set of functions in W_h that are affine on each $K \in \mathcal{T}_h$, continuous at the midpoint of each edge $e_{K-K'}$ and attain at the midpoint of each edge of Γ the value of the average of g on that edge. The solution u_h computed by the primal hybrid method is coincident with the solution u_h^{NC} computed by the non-conforming method in the sense that the value of u_h at each edge midpoint

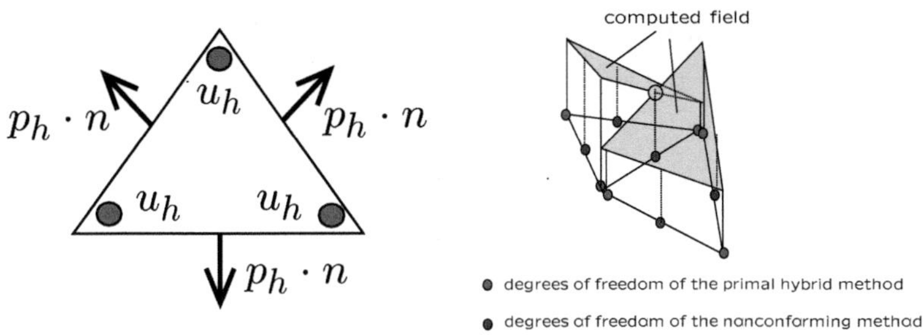

Figure 5. Degrees of freedom for the PH (left) and NC (right) formulations.

coincides with the value of u_h^{NC} at the same location. The values attained by u_h at the edge midpoints from each one of the two triangles sharing that edge are coincident (up to machine precision). The primal-hybrid method enjoys the following properties:

- the hybrid variable $p_h \cdot n$ can be eliminated by static condensation in favor of the sole unknown u_h. This leads to solving a linear algebraic system for u_h^{NC} that is symmetric and positive definite and has dimension equal to the number of internal edges of $\mathcal{E}_h$. The resulting primal-hybrid formulation can be regarded as a generalized (nonconforming) displacement-based finite element method;

- the variable u_h^{NC} satisfies the optimal error estimate

$$||u - u_h^{NC}||_{0,\Omega} \leq Ch^2|u|_{2,\Omega};$$

- starting from $p_h \cdot n$, it is possible to reconstruct an approximate stress field which is self–equilibrated within each element and satisfies traction reciprocity across each internal edge of $\mathcal{T}_h$, i.e.

$$p_h^K \cdot n_K + p_h^{K'} \cdot n_{K'} = 0 \qquad \forall e_{K-K'} \in \mathcal{T}_h. \tag{2}$$

This latter relation expresses the fact that the action–reaction principle is satisfied.

3.3 Experimental Convergence Analysis of PH Method

Let us consider the following elliptic boundary value model problem

$$\begin{cases} -\mathrm{div}\,(\nabla C) = -x\exp(y) & \text{in } \Omega, \\ C = x & \text{on } \Gamma_{D_1}, \\ C = 0 & \text{on } \Gamma_{D_2}, \\ \dfrac{\partial C}{\partial n} = x\exp(3) & \text{on } \Gamma_N, \\ \dfrac{\partial C}{\partial n} - \dfrac{1}{4}C = 0 & \text{on } \Gamma_R, \end{cases}$$

with exact solution $C(x,y) = x\exp(y)$. Figs. 7–8 show the validation of

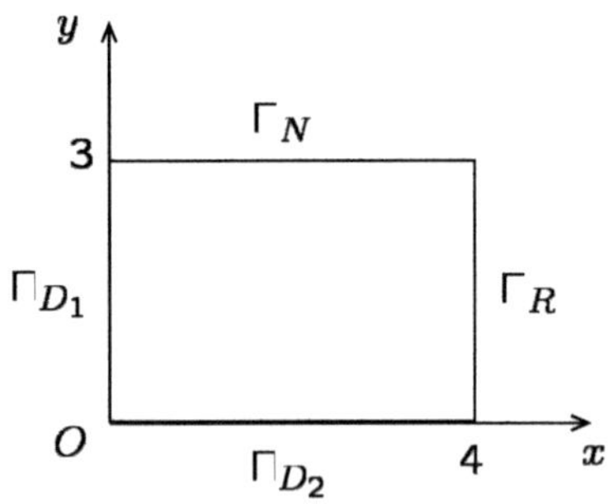

Figure 6. The computational domain.

the numerical performance of the PH–NC method discussed in the previous section. The results agree with the theoretical estimates; moreover, if a structured mesh is adopted, experiments reveal that $||p \cdot n - p_h \cdot n||_{-1/2,h} \leq Ch^2$ (cf. [3] for definition of the norm).

4 The Diffusion-Reaction Problem

In this section, we briefly describe the mathematical model of the diffusion–reaction step in the thermal oxidation process. For computational convenience, the differential problem is studied in an half part of the sole oxide domain Ω (cf. Fig. 9) and consists of finding the oxidant concentration C

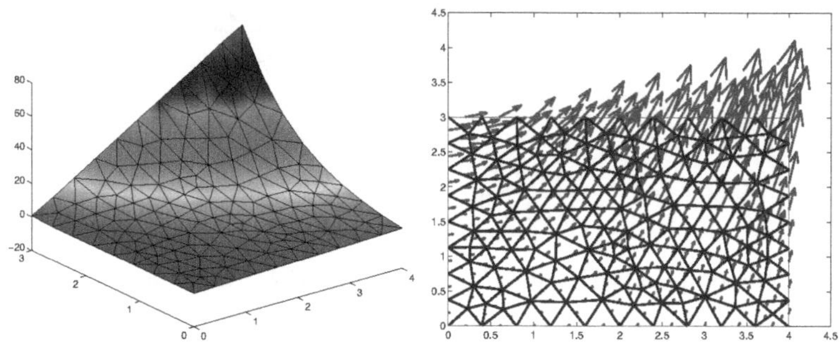

Figure 7. 3D plot of C_h (left) and vector plot of the flux interpolated using a $\mathbb{RT}_0$ reconstruction of $p_h \cdot n$ over $\mathcal{T}_h$ (right).

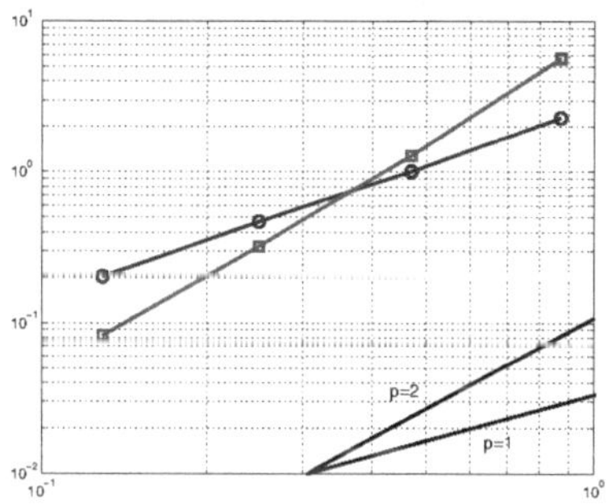

Figure 8. Error curves: $||C - C_h||_{0,\Omega}$ ('□'), $||p \cdot n - p_h \cdot n||_{-1/2,h}$, ('o').

in the oxide domain Ω such that

$$\begin{cases} -\mathrm{div}\,(D\nabla C) = 0 & \text{in} \quad \Omega \\ C = C^* & \text{on} \quad \Gamma_D, \\ D\dfrac{\partial C}{\partial n} + k_s C = 0 & \text{on} \quad \Gamma_R, \\ D\dfrac{\partial C}{\partial n} = 0 & \text{on} \quad \Gamma_N, \\ D = D_0 \exp\left(\dfrac{pV_d}{K_B T}\right), & k_s = k_{s_0} \exp\left(\dfrac{\sigma_{nn} V_r}{K_B T}\right). \end{cases} \tag{3}$$

Notice that the diffusion coefficient D and the reaction k_s coefficient are strongly nonlinear functions of the stress field. Problem (3) is numerically approximated using a primal hybrid method implemented as a nonconforming method, that reads:

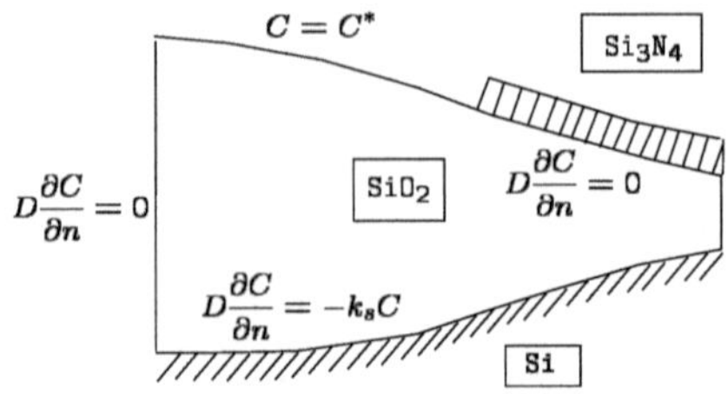

Figure 9. Computational domain for the diffusion–reaction problem.

find $C_h^{NC} \in W_{h,C^*}^{NC}$ such that

$$\sum_{K\in\mathcal{T}_h}\left(\int_K D\nabla C_h^{NC}\cdot\nabla v_h\,dx+\int_{\partial K\cap\Gamma_R} D^{-1}k_s\,C_h^{NC}\,v_h\,ds\right)=0 \qquad \forall v_h\in W_{h,0}^{NC}.$$

4.1 Diffusion-Reaction in Realistic Oxidation Simulation

To test the diffusion reaction problem solver, we study in realistic conditions the diffusion of oxygen in three layers of oxide each $1.5\cdot 10^{-4}$ cm wide and with thicknesses: $1.5\cdot 10^{-7}$ cm, $1.5\cdot 10^{-5}$ cm and $1.5\cdot 10^{-4}$ cm, respectively. In all the numerical experiments we have set $k_s = -7.8752\cdot 10^{-3}$ cm/s, $C^* = 5\cdot 10^{16}$ cm^{-3} and $D = 1.3876\cdot 10^{-6}$ cm^{2}s^{-1}. Results show

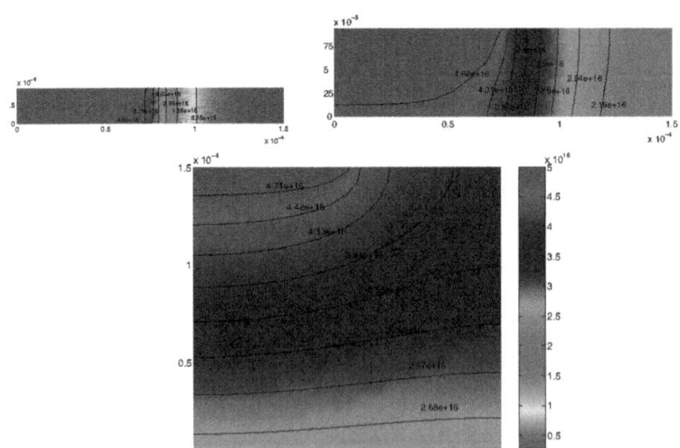

Figure 10. Simulation of diffusion process in oxide structures with varying thickness.

that oxygen distribution is practically uniform across the thickness of the oxide when the oxide layer is very thin, while it varies more considerably when the oxide layer has grown. This behavior agrees with the evidence that the slow process is the chemical reaction when the oxide thickness is small, while the slow process is the diffusion when the oxide is thicker.

5 Dual–Mixed Formulations

In this section we consider a formulation of problem (1), which provides a direct approximation of the stress field. Again, for further information we refer to [18, 2, 19, 3].

Let $\sigma = \nabla u$ (stress) be the dual unknown associated with the primal variable u (displacement). Then, the mixed strong form of (1) is the following first–order system

$$\begin{cases} -\operatorname{div}\sigma = f & in\ \Omega, \\ \sigma = \nabla u & in\ \Omega, \\ u = 0 & on\ \Gamma. \end{cases}$$

Let us introduce the functional spaces

$$\Sigma = H(\operatorname{div};\Omega), \qquad V = L^2(\Omega).$$

Then, the dual mixed (DM) weak formulation of the elliptic model problem (1) reads:
find $(\sigma, u) \in (\Sigma \times V)$ such that

$$\begin{cases} \displaystyle\int_\Omega \sigma\cdot\tau\, dx + \int_\Omega u \operatorname{div}\tau\, dx = 0 & \forall\tau\in\Sigma, \\ \displaystyle\int_\Omega (\operatorname{div}\sigma + f)\, v\, dx = 0 & \forall v \in V. \end{cases}$$

The mixed variable u is the Lagrange multiplier which enforces the equilibrium constraint

$$\operatorname{div}\sigma + f = 0 \qquad \text{in } \Omega.$$

5.1 Approximation Properties of DM Formulation

Let uis introduce the finite dimensional spaces

$$\Sigma_h = \left\{\tau_h \in \Sigma \,|\, \tau_h^K \in \mathbb{RT}_0(K)\, \forall K \in \mathcal{T}_h\right\},$$
$$V_h = \left\{v_h \in V \,|\, v_h^K \in \mathbb{P}_0(K)\, \forall K \in \mathcal{T}_h\right\}.$$

Using the abstract Babuska–Brezzi error analysis and standard interpolation results, we have the optimal error estimate

$$||\sigma - \sigma_h||_{\mathrm{div},\Omega} + ||u - u_h||_{0,\Omega} \leq Ch(|\sigma|_{1,\Omega} + |\mathrm{div}\,\sigma|_{1,\Omega} + |u|_{1,\Omega}).$$

The following properties are satisfied:

- the stress field σ_h is self-equilibrated element-by-element and satisfies traction reciprocity (2) across each internal edge of $\mathcal{T}_h$;
- the displacement u_h is such that

$$||P_h u - u_h||_{0,\Omega} \leq Ch^2 |u|_{3,\Omega},$$

where P_h is the L^2-projection onto piecewise constant functions over $\mathcal{T}_h$.

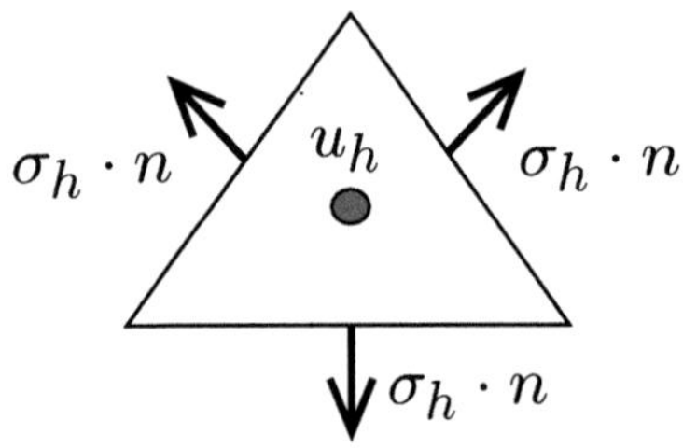

Figure 11. Degrees of freedom of the DM method.

5.2 Algebraic Form of DM Method

The discrete DM formulation in compact notation is:
find $(\sigma_h, u_h) \in (\Sigma_h \times V_h)$ such that

$$\begin{cases} a(\sigma_h, \tau_h) + b(u_h, \tau_h) = 0 & \forall \tau_h \in \Sigma_h, \\ b(v_h, \sigma_h) = (f, v_h) & \forall q_h \in V_h. \end{cases}$$

The previous discrete saddle-point problem can be written in matrix form as

$$\begin{bmatrix} A & B^T \\ B & O \end{bmatrix} \begin{pmatrix} \boldsymbol{\sigma} \\ \mathbf{u} \end{pmatrix} = \begin{pmatrix} \mathbf{0} \\ \mathbf{f} \end{pmatrix}. \tag{4}$$

We notice that the dimension of system (4) is $N_h = \dim(\Sigma_h) + \dim(V_h)$. The coefficient matrix is symmetric but, in general, indefinite; in particular, matrix A is symmetric and positive definite, but A^{-1} is full (it is impossible to easily eliminate $\boldsymbol{\sigma}$!). We refer the reader to Prof. Brenner's lectures for an efficient solution of system (4).

5.3 The DM-Hybrid Formulation

The previous drawbacks of DM formulations can be overcome by relaxing the constraint $\sigma \in H(\mathrm{div}\,;\Omega)$ through the use of a Lagrangian multiplier λ (hybrid variable). With this aim, let p, q be conjugate exponents, with $4/3 < p < 2$, and introduce the broken spaces (see [9, 5])

$$
\begin{aligned}
\widetilde{\Sigma} &= \prod_{K\in\mathcal{T}_h} W_q(\mathrm{div}\,;K),\\
\Lambda &= \left\{\lambda \in \prod_{K\in\mathcal{T}_h} W^{1/q,p}(\partial K)\,|\,\lambda_K = \lambda_{K'} \text{ on } e_{K-K'} \forall K\in\mathcal{T}_h,\ \lambda = 0 \text{ on } \Gamma\right\}.
\end{aligned}
$$

Since $p < 2$, we can use discontinuous piecewise finite elements for the approximation of functions in Λ, at the expense of a slightly stronger regularity on functions belonging to $W_q(\mathrm{div}\,;K)$ (indeed we have $q > 2$, and we can still use, in practice, the same finite elements valid for the case $q = 2$, i.e., $H(\mathrm{div}\,;K)$!!). Then, the dual-mixed hybrid (DMH) weak formulation of (1) reads:
find $(\sigma, u, \lambda) \in (\widetilde{\Sigma} \times V \times \Lambda)$ such that

$$
\begin{cases}
\displaystyle\sum_{K\in\mathcal{T}_h}\left(\int_K \sigma\cdot\tau\,dx + \int_K u\,\mathrm{div}\,\sigma\,dx - \int_{\partial K}\lambda\,\tau\cdot n\,ds\right) = 0 & \forall \tau\in\widetilde{\Sigma},\\
\displaystyle\sum_{K\in\mathcal{T}_h}\int_K v\,\mathrm{div}\,\sigma\,dx = \int_\Omega f\,v\,dx & \forall v\in V,\\
\displaystyle -\sum_{K\in\mathcal{T}_h}\int_{\partial K}\mu\,\sigma\cdot n\,ds = 0 & \forall \mu\in\Lambda.
\end{cases}
$$

The hybrid variable λ is the Lagrange multiplier which enforces the compatibility constraint (2) across each internal edge of $\mathcal{T}_h$, and the above discussed procedure is known in the literature as hybridization (see [1] and, for recent results, [7, 8]). We set

$$
\begin{aligned}
\widetilde{\Sigma}_h &= \left\{\tau_h \in \widetilde{\Sigma}\,|\,\tau_h^K \in \mathbb{RT}_0(K)\,\forall K\in\mathcal{T}_h\right\},\\
V_h &= \left\{v_h \in V\,|\,v_h^K \in \mathbb{P}_0(K)\,\forall K\in\mathcal{T}_h\right\},\\
\Lambda_h &= \left\{\lambda_h \in \Lambda\,|\,\lambda^K \in R_0(\partial K)\,\forall K\in\mathcal{T}_h\right\},
\end{aligned}
$$

where $R_0(\partial K)$ is the space of constant functions on each edge of ∂K. Using the abstract Babuska–Brezzi error analysis and standard interpolation results, we have the optimal error estimate

$$
\begin{aligned}
&||u-u_h||_{0,\Omega} + ||\sigma-\sigma_h||_{0,\Omega} \le Ch(|u|_{1,\Omega} + |\sigma|_{1,\Omega}),\\
&||\rho_h^0\lambda - \lambda_h||_{-1/2,h} \le Ch^{2/p}(|u|_{1,\Omega} + |\sigma|_{1,\Omega} + |\mathrm{div}\,\sigma|_{1,\Omega}),
\end{aligned}
$$

where ρ_h^0 is the L^2–projection over $R_0(\partial K)$ for all $K\in\mathcal{T}_h$.

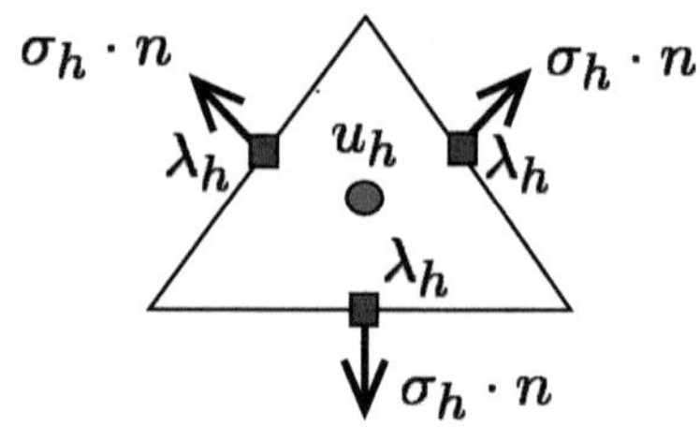

Figure 12. Degrees of freedom of DMH method.

5.4 Algebraic Form of DMH Method

The matrix form of the DMH method is

$$\begin{cases} A\boldsymbol{\sigma} + B^T\mathbf{u} + C^T\boldsymbol{\lambda} & = & \mathbf{0} \\ B\boldsymbol{\sigma} & = & \mathbf{f} \\ C\boldsymbol{\sigma} & = & \mathbf{0}. \end{cases} \tag{5}$$

We notice that the number of unknowns is 3NE + NE + Ni, where NE is the number of mesh elements and Ni is the number of internal edges. Moreover, unlike the case of DM method, the variable σ_h is fully discontinuous over $\mathcal{T}_h$, so that it can be eliminated element-by-element by static condensation (as well as u_h), to obtain

$$\boldsymbol{\sigma} = -A^{-1}\left(B^T\mathbf{u} + C^T\boldsymbol{\lambda}\right) \qquad \forall K \in \mathcal{T}_h$$
$$\mathbf{u} = \left(BA^{-1}B^T\right)^{-1}\left(\mathbf{f} - BA^{-1}C^T\boldsymbol{\lambda}\right) \qquad \forall K \in \mathcal{T}_h.$$

Replacing the above relations into (5), we get the following linear algebraic system in the sole unknown λ_h, which constitutes the DMH approximation of problem (1)

$$M_\lambda \boldsymbol{\lambda} = \mathbf{g} \tag{6}$$

where

$$M_\lambda = CA^{-1}C^T + CA^{-1}B^T(BA^{-1}B^T)^{-1}BA^{-1}C^T$$
$$\mathbf{g}_\lambda = -CA^{-1}B^T(BA^{-1}B^T)^{-1}\mathbf{f}.$$

The DMH formulation is characterized by the following properties:

- the variable λ_h enjoys a superconvergence rate because $4/3 < p < 2$;
- the stress field σ_h is self-equilibrated within each element and satisfies (2);
- the variables σ_h and u_h (post)-computed by the DMH method coincide with those of the DM method;

- matrix M_λ is symmetric and positive definite, as in the case of the standard Galerkin displacement-based formulation.

The DMH method with hybridization can be regarded as a generalized (nonconforming) displacement-based finite element approximation of (1), which gathers in a rather complete sense all the pros of the dual mixed and hybrid formulations, at the expense of a slightly greater computational cost if compared with standard primal displacement methods (that however fail at satisfying both self-equilibrium and interelement traction reciprocity).

6 The Fluid-Mechanical Problem

In this section, we briefly describe the mathematical model of the fluid–mechanical step in the thermal oxidation process. The Si_3N_4 stripe and the Si bulk can be modeled - in a first basic implementation - as linear elastic materials (with anisotropic elastic properties in the case of Si), while the SiO_2 can be represented by an incompressible fluid with nonlinear viscosity of Eyring type

$$\mu(\sigma) = \mu_0 \frac{\tau_{max}/\sigma_c}{\sinh(\tau_{max}/\sigma_c)}, \qquad \tau_{max} = \sqrt{(\sigma_{11} - \sigma_{22})^2/4 + \sigma_{12}^2},$$

where μ_0 and σ_c are suitable parameters defined to fit experimental data [15]. Summarizing, the mechanical evolution of all of the three physical subsystems involved in the thermal oxidation process is governed by the Lamè elasticity system supplied with appropriate description of material reology. We adopt for discretization of the fluid-mechanical problem the DMH method discussed in previous sections, because it is capable of dealing with both compressible and incompressible regimes, as happens in the case at hand.

6.1 The Incompressible Regime

Many industrial–relevant materials are nearly–incompressible (rubber), i.e., a high energy must be spent to modify their volume. Equivalently, a nearly–incompressible material subject to a stress tends to deform without an appreciable volume change. Mathematically speaking, the incompressibility condition is

$$\mathrm{Tr}\,\epsilon(u) = \frac{\Delta V}{V} = \mathrm{div}\ u = 0$$

where $\epsilon(u)$ is the infinitesimal strain tensor and $\Delta V/V$ is volumetric deformation. Denote by $\widehat{\lambda}$ and $\widehat{\mu}$ the Lamé parameters and let

$$G = \widehat{\mu} = \frac{E}{2(1+\nu)}, \qquad K = \widehat{\lambda} + \frac{2}{3}\widehat{\mu} = \frac{E\nu}{(1+\nu)(1-2\nu)} + \frac{2}{3}\widehat{\mu},$$
$$\rho := \frac{K}{G} = \frac{2(1+\nu)}{3(1-2\nu)}.$$

Then, it follows that

$$\operatorname{Tr}\epsilon(u) = 0 \Leftrightarrow \rho \to +\infty \Leftrightarrow \nu \to \frac{1}{2} \Leftrightarrow \widehat{\lambda} \to +\infty.$$

Large values of the Lamé parameter $\widehat{\lambda}$ make the numerical solution of the elasticity problem quite critical, and the convergence of the standard Galerkin finite element method may become quite poor even for fine grids, as shown in Fig. 13. This degradation of accuracy is referred to as volumetric locking (see [13] for a detailed discussion of this issue).

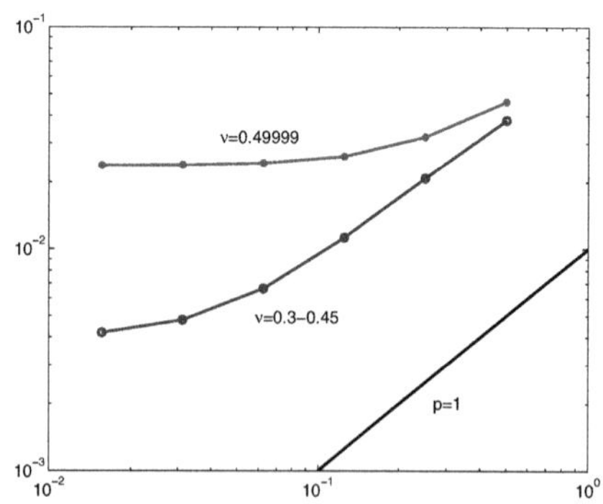

Figure 13. The locking phenomenon.

6.2 The Linear Elasticity System in Mixed Form

To overcome the previous difficulty, we consider the following form of the linear elasticity system [3, 5, 6]: find σ, u, p and γ such that

$$
\begin{cases}
\operatorname{div}\sigma = -f & \textit{in } \Omega, \\
\sigma = 2\widehat{\mu}\underbrace{(\nabla u - \gamma\chi)}_{\varepsilon(u)} - \dfrac{\widehat{\lambda}}{(\widehat{\lambda}+\widehat{\mu})}p\delta & \textit{in } \Omega, \\
p = -\dfrac{1}{2}\operatorname{Tr}\sigma & \textit{in } \Omega, \\
\operatorname{As}\sigma = 0 & \textit{in } \Omega, \\
u = 0 & \textit{on } \Gamma,
\end{cases}
\tag{7}
$$

where homogeneous Dirichlet boundary conditions have been enforced only for sake of simplicity and

$$
\chi = \begin{bmatrix} 0 & -1 \\ 1 & 0 \end{bmatrix}, \quad \delta = \begin{bmatrix} 1 & 0 \\ 0 & 1 \end{bmatrix}, \quad \operatorname{As}\tau = \tau_{12} - \tau_{21}, \quad \operatorname{Tr}\tau = \tau_{11} + \tau_{22}.
$$

The following remarks are relevant concerning system (7):

- the pressure parameter p is the Lagrangian multiplier of the incompressibility condition;
- introducing p extends to the DM formulation the original approach of Herrmann valid for displacement-based formulations [11, 12];
- the rotational parameter γ is the Lagrangian multiplier of the symmetry of the stress tensor;
- setting $\widehat{\lambda} = +\infty$, yields the DM formulation of the Stokes equations in fluid dynamics written in conservative form.

6.3 DMH Formulation for Linear Elasticity

The weak DMH formulation of (7) reads: find $(\sigma,\, u,\, \gamma,\, p,\, \lambda) \in (\Sigma \times V \times W \times Q \times \Lambda)$ such that

$$
\begin{cases}
a(\sigma;\tau) + b(u,\gamma,p,\lambda;\tau) = 0 & \forall\tau \in \Sigma, \\
b(v,\theta,q,\mu;\sigma) + c(p;q) = F(v) & \forall\,(v,\,\theta,\,q,\,\mu) \in (V \times W \times Q \times \Lambda)
\end{cases}
$$

where $W = L^q(\Omega)$, $Q = L^q(\Omega) \cap L^2_0(\Omega)$, and

$$a(\sigma,\tau) = \frac{1}{2\widehat{\mu}} \int_\Omega \sigma : \tau \, dx,$$

$$b(v,\theta,q;\tau) = \int_\Omega \left(v \cdot \operatorname{div} \tau + \theta \operatorname{As} \tau + \frac{\widehat{\lambda}}{2\widehat{\mu}(\widehat{\lambda}+\widehat{\mu})} q \operatorname{Tr} \tau\right) dx$$

$$c(p;q) = \int_\Omega \frac{\widehat{\lambda}}{\widehat{\mu}(\widehat{\lambda}+\widehat{\mu})} p \, q \, dx, \qquad F(v) = -\int_\Omega f \cdot v \, dx$$

and the spaces Σ, V, Λ are the vectorial counterparts of the corresponding spaces in the elliptic scalar case.

The Galerkin finite element approximation of (6.3) is: find $(\sigma_h, u_h, \gamma_h, p_h, \lambda_h) \in (\Sigma_h \times V_h \times W_h \times Q_h \times \Lambda_h)$ such that

$$\begin{cases} a(\sigma_h;\tau_h) + b(u_h,\gamma_h,p_h,\lambda_h;\tau_h) = 0 & \forall \tau_h \in \Sigma_h, \\ b(v_h,\theta_h,q_h,\mu_h;\sigma_h) + c(p_h;q_h) = F(v_h) & \forall\, (v_h, \theta_h, q_h, \mu_h) \\ & \in (V_h \times W_h \times Q_h \times \Lambda_h) \end{cases}$$

where Σ_h, V_h, W_h, Q_h and Λ_h are suitably chosen finite dimensional spaces. In particular, we notice that:

- cubic bubbles must be added for each element to the stress σ_h to stabilize the method, and nodally continuous rotations must be used to avoid rigid body motions of rotational type (see Fig. 14);
- σ_h and u_h can be eliminated via static condensation as in the scalar case;
- also p_h can be eliminated if $\widehat{\lambda} < +\infty$ (compressible case).

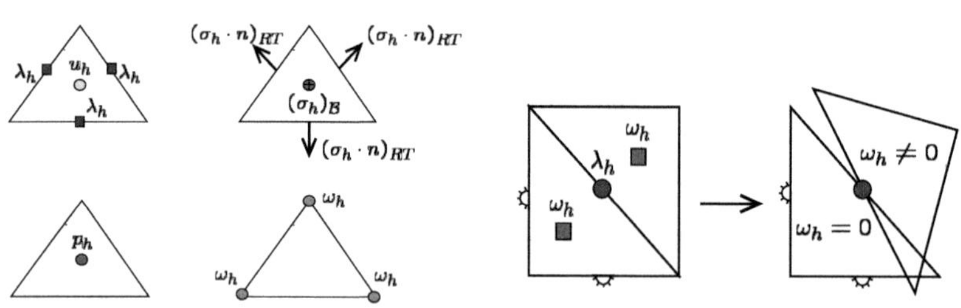

Figure 14. Degrees of freedom for the DMH approximation applied to linear elasticity (left) and mechanical explanation of the finite element space chosen for γ_h (right).

6.4 Approximation Properties of DMH Formulation

The following optimal error estimates for the mixed variables and rotation can be proved for the DMH approximation (6.3) (see [3, 5])

$$||\sigma - \sigma_h||_{0,\Omega} + ||p - p_h||_{0,\Omega} \leq Ch(|\sigma|_{1,\Omega} + |p|_{1,\Omega} + |\gamma|_{1,\Omega}),$$
$$||u - u_h||_{0,\Omega} \leq Ch(|\sigma|_{1,\Omega} + |p|_{1,\Omega} + |u|_{1,\Omega} + |\gamma|_{1,\Omega}),$$
$$||\gamma - \gamma_h||_{0,\Omega} \leq Ch(|\sigma|_{1,\Omega} + |p|_{1,\Omega} + |\gamma|_{1,\Omega}).$$

Moreover, if the triangulation is uniformly regular and $\operatorname{div}\sigma \in (H^1(\Omega))^2$, we also have

$$||P_h^0 u - u_h||_{0,\Omega} \leq Ch^2(|\sigma|_{1,\Omega} + |p|_{1,\Omega} + |\operatorname{div}\sigma|_{1,\Omega} + |\gamma|_{1,\Omega}),$$
$$||\rho_h^0 \lambda - \lambda_h||_{1/q,p,\partial K} \leq Ch^{2/p}(|\sigma|_{1,\Omega} + |p|_{1,\Omega} + |\operatorname{div}\sigma|_{1,\Omega} + |\gamma|_{1,\Omega}) \qquad \forall K \in \mathcal{T}_h.$$

Notice that, since $p < 2$, the last estimate is a superconvergence result for the hybrid variable λ_h.

6.5 Compressible material: heterogeneous elastic plate

In this section, we perform a numerical validation of the DMH formulation for linear elasticity in the case of a compressible material. A rectangular plate is considered, with a load along a portion of its upper edge with a distributed vertical force. The plate is clamped on vertical edges and is stress-free on bottom edge. The Young modulus of the materials satisfy $E_1 << E_2$, with $E_1 = 2 \cdot 1 \times 10^4$Pa and $E_2 = 2 \cdot 1 \times 10^5$Pa, respectively. Plane stress conditions are assumed. In Fig.15 we show the deformations and the stresses due to the load when the two materials are vertically stratified. Looking at the stress distribution, we clearly see that the applied load is much more consistently transferred from the material on the top to the material on the bottom when the softer material is situated over the stiffer material (Fig.15, rightmost column). This latter situation is somewhat similar to the situation occurring in the oxidation problem, where a structure made of materials with different stiffness properties is subjected to a force field.

6.6 Incompressible material: Stokes flow between two plates

In this section, we perform a numerical validation of the DMH formulation for linear elasticity in the case of an incompressible fluid squeezed between two parallel plates moving one toward the other with constant ve-

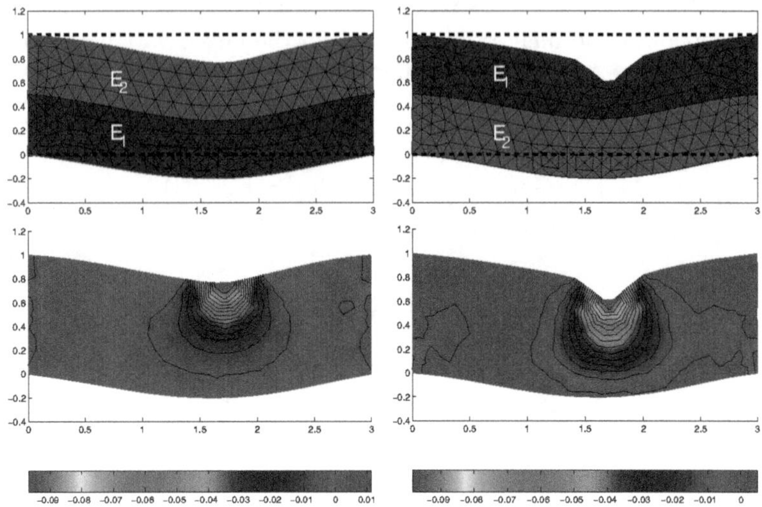

Figure 15. Deformations (amplified by a factor 100 for graphical purposes) of a rectangular plate constituted of materials with stiffness $E_2 >> E_1$ subject to a compressive load (top row) and σ_{22} component of the stress (bottom row).

locity. The exact solution is

$$\begin{cases} u = \dfrac{3V_0 x}{2b}\left(1 - \dfrac{y^2}{b^2}\right), & v = -\dfrac{V_0 y}{2b}\left(3 - \dfrac{y^2}{b^2}\right), \\[2ex] p = \dfrac{P_0}{a^2}(a^2 + y^2 - x^2), & P_0 = \dfrac{3\nu V_0 a^2}{2\,b^3}, \end{cases}$$

where ν is the kynematical viscosity and $a = 6, b = 2, V_0 = -1$. Due to the symmetry of the problem, the computation has been carried out only on a quarter of the domain, that is the dashed-line region in Fig.16. With this purpose, symmetry boundary conditions have been enforced on the left and bottom edges. The computed error curves are in agreement with the theoretical estimates given in previous section.

6.7 Thermal Oxidation Simulation

In this concluding section, we provide some numerical results concerning the complete simulation of the thermal oxidation process using the hybrid

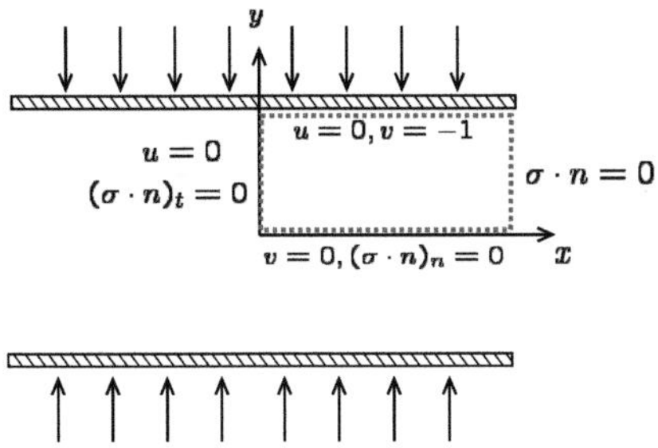

Figure 16. Geometry of the domain and boundary conditions for the Stokes problem.

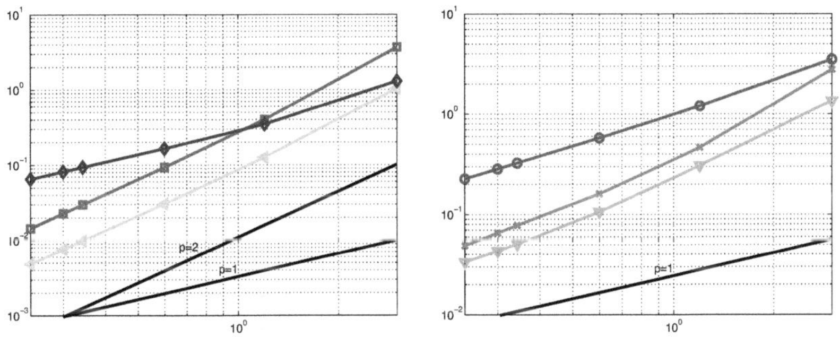

Figure 17. Error norms (left): $||u - u_h||_{(L^2(\Omega))^2}$ (diamonds), $||\rho_h^0 u - \lambda_h||_{(L^2(\mathcal{E}_h))^2}$ (squares), $||P_h u - u_h||_{(L^2(\Omega))^2}$ (triangles). Error norms (right): $||\sigma - \sigma_h||_{(L^2(\Omega))^4}$ (circles), $||p - p_h||_{L^2(\Omega)}$ (triangles), $||\gamma - \gamma_h||_{L^2(\Omega)}$ (asterisks).

and mixed–hybrid methods discussed in the lecture. A sophisticated Matlab code has been devised and tested in [6] for the study of a realistic LOCOS structure (cf. Fig. 1). Special routines have been implemented to deal with the deformation of the domain (adopting a spring-like relaxation of the grid nodes). Future release of the code will include an automatic mesh refinement procedure, driven by the a–posteriori error estimator proposed and analyzed in [4]. The computational domain is one half of the local oxidation structure shown in Fig.1 and is represented in Fig. 18. Plane strain conditions are assumed in the solution of the fluid-mechanical problem. A homogeneous Dirichlet boundary condition is enforced on the bottom of the domain, symmetry conditions hold on the vertical edges and a homogeneous Neumann condition is imposed on the remaining sides of the domain. The

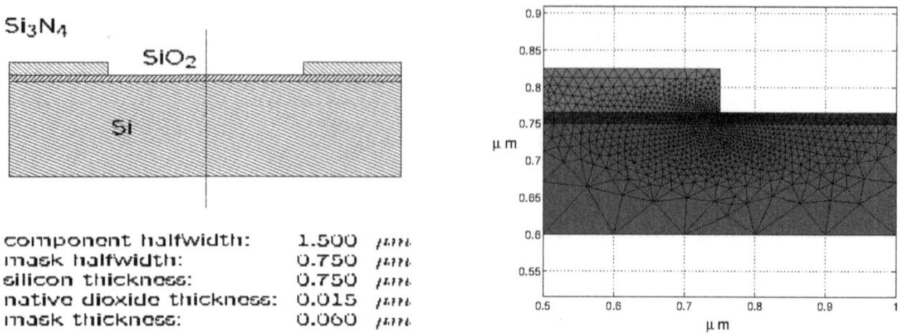

Figure 18. The silicon wafer before oxidation (left): the green color denotes the Si_3N_4 mask and the blue color denotes the native oxide film. The computational domain and its finite element discretization consisting of 2127 triangles (right).

total oxidation time considered in this simulation is 1500s. The device has semi-length 1.5 mμ, is padded with an oxide layer initially 0.015 mμ thick and is patterned with a 0.75 mμ thick nitride mask. Material properties have been chosen as in [16]. Fig. 19 shows the deformed oxide configuration after a process time evolution of 300 s. The typical "bird's beak" shape of the final oxide configuration is clearly recognizable. The largest stresses arise as expected on the junction line between the $SiO_2$and the Si_3N_4regions and in particular near the rightmost edge of the Si_3N_4band.

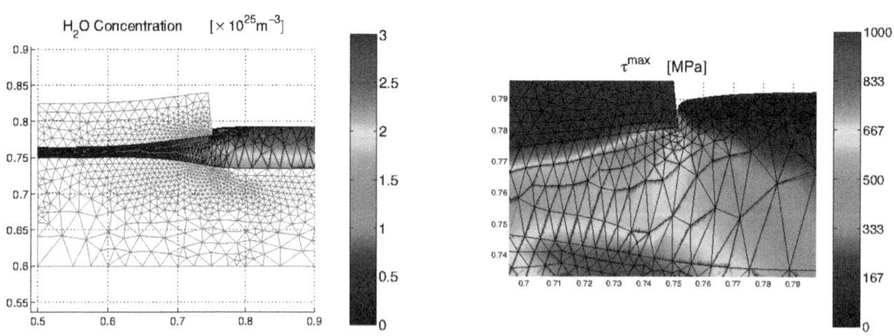

Figure 19. Deformed configuration and detailed view of the bird's beak region at $t = 300$ s.

Bibliography

[1] D. Arnold, F. Brezzi, Mixed and nonconforming finite element methods: Implementation, postprocessing and error estimates, Math. Modeling and Numer. Anal. 19-1 (1985) 7–32.

[2] F. Brezzi, M. Fortin, Mixed and Hybrid Finite Element Methods, Springer Verlag, New York, 1991.

[3] P. Causin, Mixed-Hybrid Galerkin and Petrov-Galerkin Finite Element Formulations in Fluid Mechanics, Ph.D. thesis, Università degli Studi di Milano, web–site: http://mox.polimi.it (2002).

[4] C. Carstensen, P. Causin, R. Sacco, A Posteriori Dual-Mixed (Hybrid) Adaptive Finite Element Error Control for Lamè and Stokes Equations, Numer. Math. (101), 309–332 (2005).

[5] P. Causin, R. Sacco, A dual-mixed hybrid formulation for fluid mechanics problems: mathematical analysis and application to semiconductor process technology, Comp. Meth. Appl. Mech. Engrng. (192), 593-612 (2003).

[6] P. Causin, M. Restelli, R. Sacco, A simulation system based on mixed-hybrid finite elements for thermal oxidation in semiconductor technology, Comput. Methods Appl. Mech. Engrg. (193), 3687–3710 (2004).

[7] B. Cockburn, J. Gopalakhrisnan, A characterization of hybridized mixed methods for second order elliptic problems, SIAM J. Numer. Anal. (42), 283-301 (2004).

[8] B. Cockburn, J. Gopalakhrisnan, Error analysis of variable degree mixed methods for elliptic problems via hybridization, Math. Comp. (74), 1653-1677 (2005).

[9] M. Farhloul, M. Fortin, Dual hybrid methods for the elasticity and the Stokes problems: a unified approach, Numer. Math. 76-4 (1997) 419–440.

[10] S.E. Hansen, M.D. Deal, SUPREM-IV.GS User's Reference Manual, Stanford University, 1994.

[11] L. Herrmann, Elasticity equations for incompressible and nearly - incompressible materials by a variational theorem, AIAA Jnl. 3-10 (1965) 1896–1900.

[12] L. Herrmann, R. Taylor, K. Pister, On a variational theorem for incompressible and nearly-incompressible orthotropic elasticity, Int. J. Solids Structures 4 (1968) 875–883.

[13] T.J.R. Hughes, The Finite Element Method - Linear static and dynamic finite element analysis (Cap. 4), Prentice-Hall (1987).

[14] J.P. Peng, D. Chidambarrao, G.R. Srinavasan, NOVEL. A nonlinear viscoelastic model for thermal oxidation of silicon, COMPEL, 10(4) (1991) 341–353.

[15] C. Rafferty, Stress effects in silicon oxidation - simulation and experiment, Ph.D. thesis, Stanford University (1990).

[16] V. Rao, T.J.R., Hughes, K. Garikipati, On modeling thermal oxidation in silicon. Part i-ii, Int. J. Num. Meth. Engr. 47 (2000) 341–377.

[17] P. Raviart, J. Thomas, Primal hybrid finite element methods for 2nd order elliptic equations, Math. Comp. 31-138 (1977) 391–413.

[18] P. Raviart, J. Thomas, A mixed finite element method for second order elliptic problems, in: I. Galligani, E. Magenes (Eds.), Mathematical Aspects of Finite Element Methods,I, Springer-Verlag, Berlin, 1977.

[19] J. Roberts, J. Thomas, Mixed and hybrid methods, in: P. Ciarlet, J. Lions (Eds.), Finite Element Methods, Part I, North-Holland, Amsterdam, 1991, vol.2.

Mixed Finite Element Methods -Theory and Discretization

P. Wriggers [‡]

[‡] Institute for Continuum Mechanics, Leibniz Universität Hannover

Abstract This contribution is concerned with the formulation of mixed finite elements discretization schemes for nonlinear problems of solid mechanics. Thus continuum mechanics for solids is described in the first section to provide the necessary background for the numerical method. This includes necessary kinematical relations as well as the balance laws with their weak forms and the constitutive equations. The second section then describes mixed discretization schemes which can be applied to simulate nonlinear elastic problems including finite deformations.

1 Basic Equations of Continuum Mechanics

This chapter contains a summary of the continuum mechanics background for solids which is needed for the mixed finite element formulation of solid mechanics and structural problems. The necessary kinematical relations, the balance laws with their weak forms and the constitutive equations are described in this chapter for the linear case and for finite deformations.

In the latter, kinematical relations will be formulated for the current and the referential description of motion. Based on that strain measures will be introduced. Variational formulations will be derived which are basis for nonlinear finite element methods. Isotropic hyperelastic material behaviour will be discussed as an example for nonlinear constitutive laws which can be applied to describe large strains in three-dimensional solids but also in structural elements like trusses or shells. Furthermore all relations will be provided for small strains, leading to the well known linear relations.

Since this paper is devoted to mixed nonlinear finite element formulations the underlying theory of continuum mechanics cannot be treated in the necessary depth. Due to that several extensive derivations are not presented but at such point the relevant literature is cited[1].

---

[1]The reader who needs a more in depth treatment of continuum mechanics is referred to

1.1 Kinematics

The kinematical relations concern in continuum mechanics the description of the deformation and motion of a body and the derivation of strain measures. Here a summary of the basic relation is provided.

Motion, deformation gradient In this section the motion and deformation of homogeneous bodies are considered. Here we apply the continuum approach in which a body B is described in a formal way by a set of continuously distributed points $P \in B$, also called particles or material points, which occupy a region within the Euclidean point space $\mathbb{E}^3$. A configuration of body B is a one-to-one mapping $\boldsymbol{\varphi}$: $B \longrightarrow \mathbb{E}^3$, which places the particles of B in $\mathbb{E}^3$. With this the location of a particle X from B is given for the configuration $\boldsymbol{\varphi}$ as $\boldsymbol{x} = \boldsymbol{\varphi}\,(X)$. Due to this the placement of a body B is described by $\varphi(B) = \{\,\boldsymbol{\varphi}(X)\,|\,X \in B\}$ and called configuration $\varphi(B)$ of body B.

The motion of body B is then given as a one-parametric series of configurations

$$\boldsymbol{x} = \boldsymbol{\varphi}\,(\boldsymbol{X}, t), \tag{1}$$

where $\boldsymbol{X}$ represents particle X in the reference configuration B. Based on this the placements $\boldsymbol{x}$ and $\boldsymbol{X}$ can be formulated as position vectors in $\mathbb{E}^3$ with respect to the origin $\boldsymbol{O}$, see Figure 1. Point X is defined in the reference configuration by the position vector $\boldsymbol{X} = X_A\,\boldsymbol{E}_A$. $\boldsymbol{E}_A$ defines an orthogonal base system in the reference configuration with origin $\boldsymbol{O}$. Hence one can write (1) in terms of components of the vector as

$$x_i = \varphi_i\,(X_A, t\,)\,. \tag{2}$$

If the motion is characterized with respect to the material coordinates $\{\,X_1\,, X_2\,, X_3\,\}$ we call this material or referential description. In the material description – often also referred to as Lagrangian description of motion – one follows the movement of a particle of body B in time. In the following capital letters are used as indices for components of vectors and tensors with respect to the basis $\boldsymbol{E}_A$ of the reference configuration where X_A are the Lagrangian coordinates of the particle X.

Another possibility is the use of the spatial coordinates $\{\,x_1\,, x_2\,, x_3\,\}$ when the motion of body B has to be described. In this formulation attention is paid to a point in space and the change of the motion with time

standard books, e.g. Truesdell and Toupin (1960), Truesdell and Noll (1965), Malvern (1969), Ogden (1984), Chadwick (1999) or Holzapfel (2000) or Marsden and Hughes (1983) and Ciarlet (1988) for the mathematical background of the theory of elasticity.

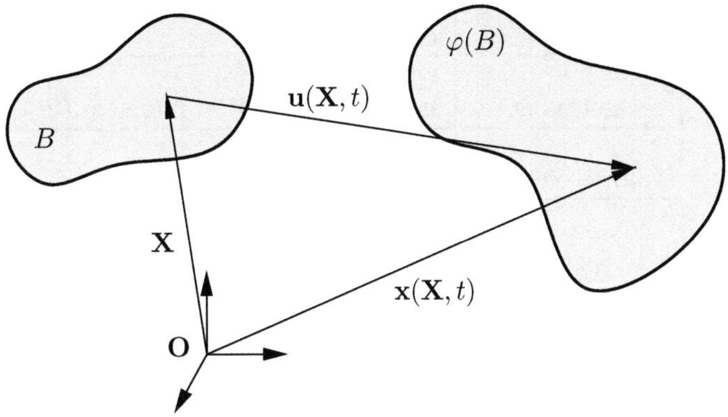

alebox1.0

Figure 1. Motion of body B

t at this point. This description is called current, spatial or Eulerian description of motion. Small letters are used for indices of vectors and tensors which are related to the basis $\boldsymbol{e}_i$ of the current or spatial configuration. The quantities x_i are the spatial coordinates of X.

By introducing a displacement vector $\boldsymbol{u}(\boldsymbol{X},t)$ as difference between the position vectors of current and initial configuration

$$\boldsymbol{u}(\boldsymbol{X},t) = \boldsymbol{\varphi}(\boldsymbol{X},t) - \boldsymbol{X} \tag{3}$$

one can express the current coordinates $\boldsymbol{x} = \boldsymbol{\varphi}\,(\boldsymbol{X},t)$ by $\boldsymbol{x} = \boldsymbol{X} + \boldsymbol{u}(\boldsymbol{X},t)$.

To be able to describe the deformation process locally the deformation gradient $\boldsymbol{F}$ is introduced which relates tangent vectors of initial and current configuration to each other. Thus the deformation gradient maps a material line element of the initial configuration $\boldsymbol{dX}$ in B, to a line element $\boldsymbol{dx}$ of the current configuration $\varphi(B)$.

$$\boldsymbol{dx} = \boldsymbol{F}\,\boldsymbol{dX}\,. \tag{4}$$

From the symbolic form $\boldsymbol{F} = \partial\,\boldsymbol{x}\,/\,\partial\,\boldsymbol{X}$ follows the components of the deformation gradient as partial derivatives $\partial x_i\,/\,\partial X_A = x_{i,A}$. With (1) and (2) on obtains

$$\boldsymbol{F} = \operatorname{Grad}\boldsymbol{\varphi}(\boldsymbol{X},t) = F_{iA}\,\boldsymbol{e}_i \otimes \boldsymbol{E}_A = \frac{\partial x_i}{\partial X_A}\,\boldsymbol{e}_i \otimes \boldsymbol{E}_A\,. \tag{5}$$

line element	$\boldsymbol{dx} = \boldsymbol{F}\,\boldsymbol{dX}$
area element	$\boldsymbol{da} = \boldsymbol{n}\,da = J\,\boldsymbol{F}^{-T}\,\boldsymbol{N}\,dA = J\,\boldsymbol{F}^{-T}\,\boldsymbol{dA}$
volume element	$dv = J\,dV$

Table 1. Transformation of differential elements

Since the gradient in (5) is a linear operator also the local transformation in (4) is linear. To maintain the connection of B during the deformation process the mapping (4) has to be one-to-one which excludes a singularity of $\boldsymbol{F}$ and further to exclude self-penetration one obtains the restriction

$$J = \det \boldsymbol{F} > 0\,, \tag{6}$$

where J defines a determinant named after Jacobi. By this $\boldsymbol{F}$ is regular and the inverse $\boldsymbol{F}^{-1}$ exists, which can be applied to invert relation (4)

$$\boldsymbol{dX} = \boldsymbol{F}^{-1}\,\boldsymbol{dx}\,. \tag{7}$$

By knowing the deformation gradient further transformations of differential quantities between B and $\varphi(B)$ can be expressed. These are the transformation of line elements, surface area elements and volume elements. The relations are summarized in Table 1

Using (3) the deformation gradient (5) follows as

$$\boldsymbol{F} = \text{Grad}\,[\,\boldsymbol{X} + \boldsymbol{u}(\boldsymbol{X},t)\,] = \boldsymbol{1} + \text{Grad}\,\boldsymbol{u} = \boldsymbol{1} + \boldsymbol{H}\,. \tag{8}$$

The tensor $\boldsymbol{H} = \text{Grad}\,\boldsymbol{u}$ is called displacement gradient.

Polar decomposition The deformation gradient can be split in a multiplicative way in a proper orthogonal rotation tensor $\boldsymbol{R}$ (with $\boldsymbol{R}^{-1} = \boldsymbol{R}^T$) and in symmetrical stretch tensors $\boldsymbol{U}\,, \boldsymbol{V}$

$$\boldsymbol{F} \;=\; \boldsymbol{R}\boldsymbol{U} = \boldsymbol{V}\,\boldsymbol{R}\,. \tag{9}$$

In general the stretch tensors $\boldsymbol{U}$ and $\boldsymbol{V}$ have to be computed via a spectral decomposition. The spectral form of $\boldsymbol{U}$ and $\boldsymbol{V}$ is given by

$$\boldsymbol{U} = \sum_{i=1}^{3} \lambda_i\,\boldsymbol{N}_i \otimes \boldsymbol{N}_i \qquad \boldsymbol{V} = \sum_{i=1}^{3} \lambda_i\,\boldsymbol{n}_i \otimes \boldsymbol{n}_i \tag{10}$$

λ_i are the principal values of the stretch tensors, also called principal stretches. They are equal for $\boldsymbol{U}$ and $\boldsymbol{V}$. The eigenvectors $\boldsymbol{N}_i$ of $\boldsymbol{U}$ are related to the reference configuration. The eigenvectors $\boldsymbol{n}_i$ of $\boldsymbol{V}$ are referred to the spatial configuration.

Strain measures In this section different strain measures are provided which are used in forthcoming formulations. The strain measures have to be objective and are not allowed to produce strains under rigid body motions. Due to that the deformation gradient itself cannot be used as strain measure. Instead one has to introduce tensors which eliminate ridgid body rotations such as $\boldsymbol{U}$ and $\boldsymbol{V}$ or the right

$$\boldsymbol{C} = \boldsymbol{F}^T \boldsymbol{F} = \boldsymbol{U}^T \boldsymbol{R}^T \boldsymbol{R} \boldsymbol{U} = \boldsymbol{U}^2 \tag{11}$$

which is related to the initial configuration B, or the left Cauchy-Green strain tensor

$$\boldsymbol{b} = \boldsymbol{F} \boldsymbol{F}^T = \boldsymbol{V} \boldsymbol{R} \boldsymbol{R}^T \boldsymbol{V}^T = \boldsymbol{V}^2 \tag{12}$$

living in the current configuration $\varphi(B)$.

Based on this strain measures the different strain tensors can be defined which are presented in Table 2

Since $\boldsymbol{C} = \boldsymbol{U}^2$ it can be easily shown that the spectral decomposition of the right Cauchy-Green tensor is given by

$$\boldsymbol{C} = \sum_{i=1}^{3} \lambda_i^2 \, \boldsymbol{N}_i \otimes \boldsymbol{N}_i \, . \tag{13}$$

Thus the generalized strain measures can be written based on the spectral decomposition (13) of the stretch tensors $\boldsymbol{U}$ and $\boldsymbol{V}$ as

For motions which are constraint by special conditions it is often possible to incorporate these constraint conditions directly into the kinematical relations. In case of incompressibility which plays a prominent role in rubber materials and metal plasticity one has the constraint condition: det $\boldsymbol{F} = J = 1$. The following multiplicative decomposition of the deformation gradient

$$\boldsymbol{F} = J^{\frac{1}{3}} \, \widehat{\boldsymbol{F}} \qquad\qquad \widehat{\boldsymbol{F}} = J^{-\frac{1}{3}} \, \boldsymbol{F} \tag{14}$$

was suggested in Flory (1961). It preserves a priori the volume of $\widehat{\boldsymbol{F}}$ (isochoric motion), since det $\widehat{\boldsymbol{F}} \equiv 1$.

By inserting (14) in (11) one obtains a relation between the isochoric part of the right Cauchy Green deformation tensor $\widehat{\boldsymbol{C}}$ and $\boldsymbol{C}$

$$\widehat{\boldsymbol{C}} = \widehat{\boldsymbol{F}}^T \, \widehat{\boldsymbol{F}} = J^{-\frac{2}{3}} \, \boldsymbol{F}^T \, \boldsymbol{F} = J^{-\frac{2}{3}} \, \boldsymbol{C} \, . \tag{15}$$

Green-Lagrange	$\boldsymbol{E} := 1/2(\boldsymbol{F}^T\boldsymbol{F} - \mathbf{1}) = \frac{1}{2}(\boldsymbol{C} - \mathbf{1})$, in B
	$\boldsymbol{E} = 1/2(F_{iA}F_{iB} - \delta_{AB})\boldsymbol{E}_A \otimes \boldsymbol{E}_B$, in B
	$\boldsymbol{E} = 1/2(\boldsymbol{H} + \boldsymbol{H}^T + \boldsymbol{H}^T\boldsymbol{H})$, in B
linear	$\boldsymbol{\varepsilon} = 1/2(\boldsymbol{H} + \boldsymbol{H}^T) = 1/2(u_{A,B} + u_{B,A})\boldsymbol{E}_A \otimes \boldsymbol{E}_B$
generalized	$\boldsymbol{E}^\alpha = 1/\alpha(\boldsymbol{U}^\alpha - \mathbf{1}), \quad \alpha \in \mathbb{R}$, in B
	$\boldsymbol{E}^\alpha = 1/\alpha\sum_{i=1}^3(\lambda_i^\alpha - 1)\boldsymbol{N}_i \otimes \boldsymbol{N}_i$, in B
	$\boldsymbol{e}^\alpha = 1/\alpha\sum_{i=1}^3(\lambda_i^\alpha - 1)\boldsymbol{n}_i \otimes \boldsymbol{n}_i \quad \alpha \in \mathbb{R}$, in $\varphi(B)$
Almansi	$\boldsymbol{e} = 1/2(\mathbf{1} - \boldsymbol{V}^{-2}) = 1/2(\mathbf{1} - \boldsymbol{b}^{-1})$, in $\varphi(B)$
Hencky	$\boldsymbol{E}^{(0)} = \ln\boldsymbol{U} \quad \text{for} \quad \alpha = 0$, in B

Table 2. Strain tensors

The multiplicative split of $\boldsymbol{F}$ in a volume changing and a volume preserving part in the nonlinear theory corresponds to an additive decomposition of the strain tensor in the geometrically linear theory in a deviator $\boldsymbol{e}_D$ and a volumetric part

$$\boldsymbol{\epsilon} = \boldsymbol{e}_D + \frac{1}{3}\,\mathrm{tr}\,\boldsymbol{\epsilon}\,\mathbf{1}\,. \tag{16}$$

Using the projection tensor $\mathbb{P} = \mathbb{I} - \frac{1}{3}\mathbf{1} \otimes \mathbf{1}$ one obtains for the deviatoric strain

$$\boldsymbol{e}_D = \mathbb{P}[\boldsymbol{\epsilon}] \tag{17}$$

1.2 Balance Equations

This section contains the differential equations which describe the local balance equations such as balance of mass, balance of linear and angular momentum.

Balance of mass We assume here only processes in which the mass of the system is conserved. This means that the change of mass has to be zero ($\dot{m} = 0$). Hence an infinitesimal mass element in initial and current configuration has to be equal which leads with $dm(\boldsymbol{X}) = dm(\boldsymbol{x})$ to $\rho\, dv = \rho_0\, dV$. Here ρ_0 and ρ are the densities in initial and current configuration, respectively. With $dv = J\, dV$ the the Lagrangian description of the mass balance is obtained

$$\rho_0 = J\, \rho\,. \tag{18}$$

For completeness we will present the rate form of mass continuity with respect to the spatial configuration

$$\dot{\rho}(\boldsymbol{x}\,, t) + \rho(\boldsymbol{x}\,, t)\,\mathrm{div}\,\boldsymbol{v}(\boldsymbol{x}\,, t) = 0 \tag{19}$$

which follows from the evaluation of $\dot{m} = \frac{D}{Dt} \int\limits_B \rho(\boldsymbol{x}\,, t)\, dv = 0\,.$

Balance of linear and angular momentum The linear momentum or the translational momentum is given in the current and initial configuration with (18) by

$$\boldsymbol{L} = \int\limits_{\varphi(B)} \rho\, \boldsymbol{v}\, dv = \int\limits_{B} \rho_0\, \boldsymbol{v}\, dV \tag{20}$$

for the continuous case. The balance of linear momentum is now expressed by

$$\dot{\boldsymbol{L}} = \int\limits_{\varphi(B)} \rho\, \bar{\boldsymbol{b}}\, dv + \int\limits_{\varphi(\partial B)} \boldsymbol{t}\, da\,. \tag{21}$$

$\rho\, \bar{\boldsymbol{b}}$ defines the volume force (e.g. gravitational force). $\boldsymbol{t}$ is the stress vector acting on the surface of the body. With Cauchy's theorem which relates the stress vector $\boldsymbol{t}$ to the surface normal $\boldsymbol{n}$ via the linear mapping

$$\boldsymbol{t} = \boldsymbol{\sigma}\, \boldsymbol{n}\,, \tag{22}$$

which can be presented in summation notation as $t_i = \sigma_{ik}\, n_k$. With the help of the divergence theorem the local balance equation of linear momentum is derived from (21) with reference to the current configuration $\varphi(B)$

$$\mathrm{div}\, \boldsymbol{\sigma} + \rho\, \bar{\boldsymbol{b}} = \rho \dot{\boldsymbol{v}}\,. \tag{23}$$

The stress tensor $\boldsymbol{\sigma}$ is called Cauchy stress tensor. Again index notation leads to $\sigma_{ik,i} + \rho\, \bar{b}_k = \rho\, \dot{v}_k$. The right hand side term $\rho \dot{\boldsymbol{v}}$ describes the inertial forces which can be neglected in case of purely static investigations.

The angular momentum with reference to a point O given by $\boldsymbol{x}_0$ is defined with respect to the current and initial configuration with (18) as

$$\boldsymbol{J} = \int\limits_{\varphi(B)} (\boldsymbol{\varphi} - \boldsymbol{x}_0) \times \rho\, \boldsymbol{v}\, dv = \int\limits_{B} (\boldsymbol{\varphi} - \boldsymbol{x}_0) \times \rho_0\, \boldsymbol{v}\, dV\,. \tag{24}$$

The balance of angular momentum can then be written as

$$\dot{\boldsymbol{J}} = \int\limits_{\varphi(B)} (\boldsymbol{\varphi} - \boldsymbol{x}_0) \times \rho\, \bar{\boldsymbol{b}}\, dv + \int\limits_{\varphi(\partial B)} (\boldsymbol{\varphi} - \boldsymbol{x}_0) \times \boldsymbol{t}\, da\,. \tag{25}$$

This equation yields after some manipulations the local balance of angular momentum which simply demands the symmetry of the Cauchy stress tensor

$$\boldsymbol{\sigma} = \boldsymbol{\sigma}^T\,, \qquad \text{or} \quad \sigma_{ik} = \sigma_{ki}\,. \tag{26}$$

Introduction of different stress tensors Equations (23) and (26) are referred to the current configuration. Often it is desirable to relate all quantities to the initial configuration B. Since a given stress vector does not change when referred to the current or initial configuration the following transformation can be derived by using Nanson's formula

$$\int\limits_{\partial\varphi(B)} \boldsymbol{\sigma}\, \boldsymbol{n}\, da = \int\limits_{\partial B} \boldsymbol{\sigma}\, J\, \boldsymbol{F}^{-T}\, \boldsymbol{N}\, dA = \int\limits_{\partial B} \boldsymbol{P}\, \boldsymbol{N}\, dA\,. \tag{27}$$

This relation defines the first Piola-Kirchhoff stress tensor $\boldsymbol{P}$. Different other stress measures can be defined which can be found in Table 3.

Since it is simpler to work in the initial configuration the symmetrical stress tensor, the 2. Piola-Kirchhoff stress tensor, was introduced, see Table 3. $\boldsymbol{S}$ does not represent a stress which can be interpreted physically. Hence it is a pure mathematical quantity which however plays a prominent role in constitutive theory since it is work conjugate to the Green-Lagrange strain tensor, see Table 2.

Balance equations with respect to initial configuration With the first Piola-Kirchhoff stress the local balance of linear momentum (23) can be recast with respect to the initial configuration as

$$\text{DIV}\, \boldsymbol{P} + \rho_0\, \bar{\boldsymbol{b}} = \rho_0\, \dot{\boldsymbol{v}} \tag{28}$$

where DIV denotes the divergence operation with respect to the initial configuration. Furthermore the balance of angular momentum (26) yields

$$\boldsymbol{P}\, \boldsymbol{F}^T = \boldsymbol{F}\, \boldsymbol{P}^T\,. \tag{29}$$

1. Piola-Kirchoff	$\boldsymbol{P} = J\,\boldsymbol{\sigma}\boldsymbol{F}^{-T}$
	$P_{Ak} = J\,\sigma_{ik}(F_{iA})^{-1}$
2. Piola Kirchoff	$\boldsymbol{S} = \boldsymbol{F}^{-1}\,\boldsymbol{P} = J\,\boldsymbol{F}^{-1}\,\boldsymbol{\sigma}\boldsymbol{F}^{-T}$
	$S_{AB} = (F_{Ai})^{-1}\,P_{Bi} = J\,(F_{Ai})^{-1}\,\sigma_{ik}(F_{kB})^{-1}$
Kirchoff	$\boldsymbol{\tau} = \boldsymbol{F}\,\boldsymbol{S}\,\boldsymbol{F}^{T}\,, \quad \boldsymbol{\tau} = J\,\boldsymbol{\sigma}$
	$\tau_{ik} = (F_{iA})\,S_{AB}\,(F_{Bk})$

Table 3. Different stress tensors and their definitions

From this it is clear that the first Piola-Kirchhoff stress tensor is non-symmetric. Using the relations provided in Table 3 we see that balance of angular momentum yields the symmetry of the second Piola-Kirchhoff stress tensor: $\boldsymbol{S} = \boldsymbol{S}^T$.

1.3 Constitutive Equations

The kinematical relations and balance laws derived so far are not sufficient to to solve a boundary or initial value problem in continuum mechanics. For a closure and thus for a complete set of equations a formulation of a constitutive equation is needed which characterizes the material response of a body under consideration. Here only elastic material will be considered.

Elastic material Purely elastic material behaviour is discussed in this section under the assumption of hyper elasticity, see e.g. Ogden (1984). This description is valid for many materials – like e.g. foam or rubbers – which undergo finite deformations. In case of small strains these constitutive equations reduce to the classical law of Hooke known form the linear theory of elasticity.

The constitutive equation for the second Piola-Kirchhoff stress tensor can be derived from the potential ψ in case of a hyper elastic material. ψ describes the strain energy stored in the body (for this reason it is called strain energy function). The derivative of ψ with respect to the right Cauchy-

Green tensor yields[2].

$$S = 2\,\rho_0\,\frac{\partial\psi(\boldsymbol{C})}{\partial\boldsymbol{C}}\,; \qquad S_{AB} = 2\,\rho_0\,\frac{\partial\psi(C_{CD})}{\partial C_{AB}}. \tag{30}$$

The material behaviour is independent on directions for the technically important class of isotropic materials which includes e.g. steel, aluminium, rubber or concrete. With this assumption it is possible to specify the general function ψ in (30). By introducing isotropy groups a function can be derived which only depends on the invariants of the strain tensors, see e.g. Ogden (1984). For the right Cauchy-Green deformation tensor $\boldsymbol{C} = \boldsymbol{F}^T\,\boldsymbol{F}$ and the left Cauchy-Green deformation tensors $\boldsymbol{b} = \boldsymbol{F}\,\boldsymbol{F}^T$, which have due to their definition the same invariants, it follows

$$\psi(\boldsymbol{C}) = \psi\,(\,I_C,\, II_C,\, III_C\,) = \psi\,(\,I_b,\, II_b,\, III_b\,) = \psi(\boldsymbol{b}) \tag{31}$$

where I_C, II_C, III_C are the invariants. With the relations

$$\begin{aligned} I_C &= \lambda_1^2 + \lambda_2^2 + \lambda_3^2 \\ II_C &= \lambda_1^2\,\lambda_2^2 + \lambda_2^2\,\lambda_3^2 + \lambda_3^2\,\lambda_1^2 \\ III_C &= \lambda_1^2\,\lambda_2^2\,\lambda_3^2 \end{aligned} \tag{32}$$

one can express the invariants in terms of the principal stretches. Thus the strain energy function assumes the form $\psi(\boldsymbol{C}) = \psi\,(\,\lambda_1^2,\, \lambda_2^2,\, \lambda_3^2\,)$ when written in terms of the principal stretches λ_i^2 of $\boldsymbol{C}$.

This description of the material leads to the following relation between the second Piola-Kirchhoff stresses and the right Cauchy-Green tensor

$$\boldsymbol{S} = 2\,\rho_0\,\big[\,(\,\frac{\partial\psi}{\partial I_C} + I_C\,\frac{\partial\psi}{\partial II_C}\,)\,\boldsymbol{1} - \frac{\partial\psi}{\partial II_C}\,\boldsymbol{C} + III_C\,\frac{\partial\psi}{\partial III_C}\,\boldsymbol{C}^{-1}\,\big]. \tag{33}$$

In this equation the relations

$$\frac{\partial I_C}{\partial\boldsymbol{C}} = \boldsymbol{1}\,, \quad \frac{\partial II_C}{\partial\boldsymbol{C}} = I_C\,\boldsymbol{1} - \boldsymbol{C}\,, \quad \frac{\partial III_C}{\partial\boldsymbol{C}} = III_C\,\boldsymbol{C}^{-1} \tag{34}$$

were applied for the derivative of the invariant of a tensor with respect to the tensor itself.

[2] Let us remark that the exclusive dependence of ψ on $\boldsymbol{C}$ is substantiated by the general principles. For a more detailed treatment see e.g. Truesdell and Noll (1965) and Malvern (1969)

The constitutive equation (33) can be transformed to the current configuration and expressed in the Cauchy stresses in terms of the left Cauchy-Green tensor. With the relations from Table 3 and by considering (18) it follows

$$\boldsymbol{\sigma} = 2\,\rho\,\boldsymbol{F}\,\frac{\partial\psi(\boldsymbol{C})}{\partial\boldsymbol{C}}\,\boldsymbol{F}^T$$

This yields with the fact that the invariants of $\boldsymbol{C}$ and $\boldsymbol{b}$ are equal and with $\boldsymbol{F}\boldsymbol{C}^{-1}\boldsymbol{F}^T = \boldsymbol{1}$

$$\boldsymbol{\sigma} = 2\,\rho\,\big[\,(\frac{\partial\psi}{\partial I_b} + I_b\,\frac{\partial\psi}{\partial II_b}\,)\,\boldsymbol{b} - \frac{\partial\psi}{\partial II_b}\,\boldsymbol{b}^2 + III_b\,\frac{\partial\psi}{\partial III_b}\,)\,\boldsymbol{1}\,\big]\,. \tag{35}$$

By comparison with (33) the equivalence of the last relation with

$$\boldsymbol{\sigma} = 2\,\rho\,\boldsymbol{b}\,\frac{\partial\psi(\boldsymbol{b})}{\partial\boldsymbol{b}}\,.$$

can be shown. This relation can be used to compute the Cauchy stress tensor directly for an isotropic material.

To simplify notation the specific strain energy $\rho_0\,\psi$ is written as W. Now different strain energy functions and the resulting stress strain relations are summarized in Table 4. Constitutive equations of the form (33) are still very complex because ψ can be an arbitrary function of the invariants. This can lead to a large number of constitutive parameters which are not easily determined by experiments.

The St. Venant material in Table 4 corresponds to Hooke's laws of the infinitesimal theory of elasticity with the Lamé constants Λ and μ. These constants can be converted in the modulus of elasticity $E = \frac{(3\Lambda+2\mu)\,\mu}{\Lambda+\mu}$ and Poisson's ratio $\nu = \frac{\Lambda}{2\,(\Lambda+\mu)}$. Generally it is known that the St. Venant constitutive equation is restricted to deformations with finite rotations but small strains[3].

In case of geometrically linear materials a linear relation between the stresses and the linear strains is is given in terms of the two Lamé constants Λ and μ. Since in the linear case we do not have to distinguish different stress measures, in Table 4 the Cauchy stress is used for convenience.

---

[3] St. Venant's law depicts major deficiencies in the compressible range: in the limit case of the compression of a body to volume „0“the stress $\boldsymbol{\sigma}$ approaches zero instead of $\lim_{J\to 0}\boldsymbol{\sigma} \to -\infty$. With such behaviour this material equation is not applicable for general simulations in finite elasticity. However it can be successfully used for large deflection analysis of thin structural members like beams or shells.

Neo-Hooke	$W(I_C\,,J) = g(J) + \frac{1}{2}\,\mu\,(I_C - 3)$
	$g(J) = c\,(J^2 - 1) - d\ln J - \mu\ln J \quad \text{with} \quad c > 0\,, d > 0$
	$\boldsymbol{S} = \Lambda\,/\,2\,(\,J^2 - 1\,)\,\boldsymbol{C}^{-1} + \mu\,(\,\mathbf{1} - \boldsymbol{C}^{-1}\,),\, c = \frac{\Lambda}{4},\, d = \frac{\Lambda}{2}$
	$\boldsymbol{\sigma} = \Lambda\,/\,2\,(\,J - 1\,/\,J\,)\,\mathbf{1} + \mu\,/\,J\,(\,\boldsymbol{b} - \mathbf{1}\,)$
Mooney-Rivlin	$W(I_C,\, II_C) = c_1\,(I_C - 3) + c_2\,(II_C - 3)$
	$\boldsymbol{S} = 2\,c_1\,\mathbf{1} + 2\,c_2\,(\,I_C\,\mathbf{1} - \boldsymbol{C})$
Ogden	$W(\lambda_k) = \sum_{i=1}^{r}\,\mu_i\,K_i\,(\lambda_k) + g(J)\,;$
	$K_i(\lambda_k) = 1\,/\,\alpha_i\,(\lambda_1^{\alpha_i} + \lambda_2^{\alpha_i} + \lambda_3^{\alpha_i} - 3)$
St. Venant	$W(\boldsymbol{E}) = 1\,/\,2\,\Lambda\,(\mathrm{tr}\,\boldsymbol{E})^2 + \mu\,\boldsymbol{E}\cdot\boldsymbol{E})$
	$\boldsymbol{S} = \Lambda\,\mathrm{tr}\,\boldsymbol{E}\,\mathbf{1} + 2\,\mu\,\boldsymbol{E}$
linear	$W(\boldsymbol{\epsilon}) = 1\,/\,2\,\Lambda\,(\mathrm{tr}\,\boldsymbol{\epsilon})^2 + \mu\,\boldsymbol{\epsilon}\cdot\boldsymbol{\epsilon})$
	$\boldsymbol{\sigma} = \Lambda\,\mathrm{tr}\,\boldsymbol{\epsilon}\,\mathbf{1} + 2\,\mu\,\boldsymbol{\epsilon}$

Table 4. Different strain energy functions and resulting stresses.

Split in isochoric and volumetric parts, finite strains In finite elasticity of foam or rubber materials one splits of the deformation in a volumetric part represented by J and the isochoric part described by $\widehat{\boldsymbol{C}}$, see (15), since both parts can depict different material behaviour. This split is also useful in mixed finite element formulations when quasi-incompressible materials are described by special numerical formulations since the split permits a different treatment of the incompressible part.

One possibility for the formulation of the constitutive equation is given by an additive split of the strain energy function in its volumetric and isochoric parts: $W(\widehat{\boldsymbol{C}}, J) = \hat{W}(\widehat{\boldsymbol{C}}) + U(J)$. This leads for the strain energy function introduced in Table 4 to

$$W(\widehat{\boldsymbol{C}}, J) = U(J) + \frac{1}{2}\,\mu\,(I_{\widehat{C}} - 3)\,. \tag{36}$$

Here the term $U(J)$ is different from $g(J)$: $U(J) = \frac{K}{4}\,(J^2 - 1) - \frac{K}{2}\ln J$ where K denotes the modulus of compression.

The second Piola-Kirchhoff stresses are computed via the chain rule

$$\boldsymbol{S} = 2\,\frac{\partial W}{\partial \boldsymbol{C}} = 2\,\frac{\partial \hat{W}}{\partial \widehat{\boldsymbol{C}}}\,\frac{\partial \widehat{\boldsymbol{C}}}{\partial \boldsymbol{C}} + 2\,\frac{\partial U}{\partial J}\,\frac{\partial J}{\partial \boldsymbol{C}}\,. \tag{37}$$

For an explicit evaluation of this equation one needs the derivatives $\partial J \,/\, \partial \boldsymbol{C}$ and $\partial \widehat{\boldsymbol{C}} \,/\, \partial \boldsymbol{C}$. We obtain

$$\frac{\partial J}{\partial \boldsymbol{C}} = \frac{1}{2}\,J\,\boldsymbol{C}^{-1}\,, \tag{38}$$

see also $(34)_3$, and can write with (15)

$$\frac{\partial \widehat{\boldsymbol{C}}}{\partial \boldsymbol{C}} = J^{-\frac{2}{3}}\,(\mathbb{I} - \frac{1}{3}\,\boldsymbol{C}^{-1} \otimes \boldsymbol{C}\,) =: \mathbb{P} \tag{39}$$

In this equation the fourth order unit tensor $\mathbb{I}_{ABEF} = \frac{1}{2}\,(\delta_{AE}\delta_{BF} + \delta_{AF}\delta_{BE})$ follows from $\partial \boldsymbol{C} \,/\, \partial \boldsymbol{C}$. The explicit form of (37) can now be stated as

$$\boldsymbol{S} = \mathbb{P}\,[\,2\,\frac{\partial \hat{W}}{\partial \widehat{\boldsymbol{C}}}\,] + \frac{\partial U}{\partial J}\,J\,\boldsymbol{C}^{-1} = \boldsymbol{S}_{ISO} + \boldsymbol{S}_{VOL}\,. \tag{40}$$

For the special choice of the strain energy function (36) one specifies

$$\boldsymbol{S}_{ISO} = \mu\,\mathbb{P}\,[\,\mathbf{1}\,]\,, \qquad \boldsymbol{S}_{VOL} = \frac{K}{2}\,(\,J^2 - 1\,)\,\boldsymbol{C}^{-1}\,. \tag{41}$$

In explicit form one obtains for the isochoric part of the second Piola-Kirchhoff stress tensor: $\boldsymbol{S}_{ISO} = \mu\,J^{-\frac{2}{3}}\,(\mathbf{1} - \frac{1}{3}\,\mathrm{tr}\boldsymbol{C}\,\boldsymbol{C}^{-1}\,)$.

The transformation to the current configuration yields for the Kirchhoff stress tensor introduced in Table 3. By defining the operator $\mathrm{dev}(\bullet) = (\bullet) - \frac{1}{3}\,\mathrm{tr}(\bullet)\,\mathbf{1}$ one can write (40) as

$$\boldsymbol{\tau} = J\,p\,\mathbf{1} + \mathrm{dev}\,\widehat{\boldsymbol{\tau}} = \tau_{vol}\,\mathbf{1} + \boldsymbol{\tau}_{iso}\,. \tag{42}$$

This depicts the split of the stress tensor into a volumetric and an isochoric part clearly. The following definitions were used in (42)

$$p = \frac{\partial U}{\partial J} \qquad \text{and} \qquad \widehat{\boldsymbol{\tau}} = \widehat{\boldsymbol{F}}\,2\,\frac{\partial \hat{W}}{\partial \widehat{\boldsymbol{C}}}\,\widehat{\boldsymbol{F}}^T\,. \tag{43}$$

Split in isochoric and volumetric parts, small strains In case of small strains, we can use (16) to represent the linear elastic constitutive equation by

$$\boldsymbol{\sigma} = \Lambda \operatorname{tr} \boldsymbol{\epsilon}\, \mathbf{1} + 2\,\mu\,(\, \boldsymbol{e}_D + \frac{1}{3} \operatorname{tr} \boldsymbol{\epsilon}\,) \tag{44}$$

This equation can be rewritten with the modulus of compression $K = \Lambda + \frac{2}{3}\,\mu$ as

$$\boldsymbol{\sigma} = K \operatorname{tr} \boldsymbol{\epsilon}\, \mathbf{1} + 2\,\mu\, \boldsymbol{e}_D \tag{45}$$

With the introduction of the pressure

$$p = K \operatorname{tr} \boldsymbol{\epsilon} \tag{46}$$

and the deviatoric stress

$$\boldsymbol{s} = 2\,\mu\, \boldsymbol{e}_D \tag{47}$$

the stress can be decomposed as

$$\boldsymbol{\sigma} = p\, \mathbf{1} + \boldsymbol{s}\,. \tag{48}$$

Formulation with respect to principal stretches In case that the elastic strain energy is given in terms of the principal stretches $\lambda_1\,, \lambda_2\,, \lambda_3$ one computes the second Piola-Kirchhoff stresses with (30) from

$$\boldsymbol{S} = 2\,\frac{\partial W(\lambda_k)}{\partial \boldsymbol{C}} = 2 \sum_{i=1}^{3} \frac{\partial W}{\partial \lambda_i}\,\frac{\partial \lambda_i}{\partial \boldsymbol{C}}\,. \tag{49}$$

The derivative $\partial W \,/\, \lambda_i$ can be directly obtained from W when W is given as a function of the principal stretches. The partial derivative $\partial \lambda_i \,/\, \partial \boldsymbol{C}$ which occurs when the chain rule is applied in (49) can be determined, see e.g. Simo and Taylor (1991), as

$$\frac{\partial \lambda_i}{\partial \boldsymbol{C}} = \frac{1}{2\,\lambda_i}\,\boldsymbol{N}_i \otimes \boldsymbol{N}_i\,. \tag{50}$$

Thus the second Piola-Kirchhoff stresses follow as

$$\boldsymbol{S} = \sum_{i=1}^{3} \frac{1}{\lambda_{(i)}}\,\frac{\partial W}{\partial \lambda_{(i)}}\,\boldsymbol{N}_{(i)} \otimes \boldsymbol{N}_{(i)} = \sum_{i=1}^{3} S_{(i)}\,\boldsymbol{N}_{(i)} \otimes \boldsymbol{N}_{(i)}\,. \tag{51}$$

In this equation Einstein's sum convention is not used which is denoted by the bracket around the index i.

The Kirchhoff stress tensor follows with $\boldsymbol{\tau} = \boldsymbol{F}\,\boldsymbol{S}\,\boldsymbol{F}^T$ as

$$\boldsymbol{\tau} = \sum_{i=1}^{3} \lambda_{(i)} \frac{\partial W}{\partial \lambda_{(i)}} \boldsymbol{n}_{(i)} \otimes \boldsymbol{n}_{(i)} = \sum_{i=1}^{3} \tau_{(i)}\, \boldsymbol{n}_{(i)} \otimes \boldsymbol{n}_{(i)} \,. \tag{52}$$

By comparison with (51) the principal values of the Kirchhoff stresses can be related to the principal values of the second Piola-Kirchhoff stresses via $\tau_i = \lambda_i^2\, S_i$[4].

1.4 Weak Form of Equilibrium, Variational Principles

For the analysis of static nonlinear boundary value problems in continuum mechanics one has to solve a system of partial differential equations which consist of kinematical relations, local balance of momentum and the constitutive equations. The strong form of these equations is presented in the following for hyperelastic solids by two alternative descriptions. These are the description with respect to the initial and current configurations of the bodies. For the description with respect to the initial configuration B different stress measures like the first Piola–Kirchhoff stresses or the second Piola–Kirchhoff stresses can be used. This results in two different formulations

Kinematics:	$\boldsymbol{F} = \text{Grad}\boldsymbol{\varphi}$	$\boldsymbol{E} = \frac{1}{2}\,(\boldsymbol{F}^T\,\boldsymbol{F} - \boldsymbol{1})$
Equilibrium:	$\text{Div}\,\boldsymbol{P} + \rho_0\,\bar{\boldsymbol{b}} = \rho_0\,\dot{\boldsymbol{v}}$	$\text{Div}\,(\boldsymbol{F}\,\boldsymbol{S}) + \rho_0\,\bar{\boldsymbol{b}} = \rho_0\,\dot{\boldsymbol{v}}$
Constitutive equation:	$\boldsymbol{P} = \dfrac{\partial W}{\partial \boldsymbol{F}}$	$\boldsymbol{S} = \dfrac{\partial W}{\partial \boldsymbol{E}}$

Additionally the boundary conditions for the displacements have to be prescribed on ∂B_u and boundary conditions for the tractions have to be formulated on ∂B_σ which yields

$$\boldsymbol{u} = \bar{\boldsymbol{u}} \quad \text{on} \quad \partial B_u \qquad \text{and} \qquad \boldsymbol{P}\,\boldsymbol{N} = \boldsymbol{F}\,\boldsymbol{S}\,\boldsymbol{N} = \bar{\boldsymbol{t}} \quad \text{on} \quad \partial B_\sigma$$

[4] For numerical computations it is necessary to transform the constitutive equations (51) and (52) to a cartesian coordinate system. This transformation can be performed via the relations $\boldsymbol{N}_I = \boldsymbol{D}\,\boldsymbol{E}_I$ and $\boldsymbol{n}_i = \boldsymbol{D}^\varphi\,\boldsymbol{e}_i$, respectively. The transformation matrices are defined by $\boldsymbol{D} = \boldsymbol{N}_J \otimes \boldsymbol{E}_J$ and $\boldsymbol{D}^\varphi = \boldsymbol{n}_j \otimes \boldsymbol{e}_j$. Using the component form of the transformation tensors $\boldsymbol{D}$ (the $D_{IK} = \boldsymbol{E}_I \cdot \boldsymbol{N}_K$ denote the directional cosines of the eigenvectors $\boldsymbol{N}_I$ with respect to the cartesian basis $\boldsymbol{E}_I$) we write for (51)

$$\boldsymbol{S} = \sum_{i=1}^{3} S_{(i)}\, D_{(i)\,J}\, D_{(i)\,K}\, \boldsymbol{E}_J \otimes \boldsymbol{E}_K \,. \tag{53}$$

All equations stated above can be transformed to the current configuration $\varphi(B)$ where the constitutive equations are formulated in terms of the Cauchy stress tensor, $\boldsymbol{\sigma}$, and the Kirchhoff stress tensor, $\boldsymbol{\tau}$,

Kinematics:	$\boldsymbol{b} = \boldsymbol{F}\,\boldsymbol{F}^T$	
Equilibrium:	$\operatorname{div}\boldsymbol{\sigma} + \rho\,\bar{\boldsymbol{b}} = \rho\dot{\boldsymbol{v}}$	$\operatorname{div}\left(\frac{1}{J}\,\boldsymbol{\tau}\right) + \rho\,\bar{\boldsymbol{b}} = \rho\dot{\boldsymbol{v}}$
Constitutive equation:	$\boldsymbol{\sigma} = 2\,\rho\,\boldsymbol{b}\,\dfrac{\partial\psi}{\partial\boldsymbol{b}}$	$\boldsymbol{\tau} = 2\,\boldsymbol{b}\,\dfrac{\partial W}{\partial\boldsymbol{b}}$

The displacement boundary conditions are given on $\varphi(\partial B_u)$ as $\boldsymbol{u} = \bar{\boldsymbol{u}}$. For the tractions we have $\boldsymbol{\sigma n} = \hat{\boldsymbol{t}}$ on $\varphi(\partial B_\sigma)$.

The use of the finite element method is based on variational formulations of the equations, summarized above. These variational formulations will be described in the following sections based on a referential and spatial description.

Weak form of linear momentum, initial configuration The principle of virtual work is an equivalent formulation of the balance of linear and angular momentum. It is also called weak form of the partial differential equation in the mathematical literature, resulting from linear momentum. Since no further assumptions, like existence of a potential, are made, the weak form is applicable to general problems like inelastic materials, friction, non-conservative loading, etc. The derivation of the weak form starts from the equation of linear momentum ($\operatorname{Div}\boldsymbol{P} + \rho_0\,\bar{\boldsymbol{b}} = \rho_0\,\dot{\boldsymbol{v}}$) which is multiplied scalar by a vector valued function $\boldsymbol{\eta} = \{\boldsymbol{\eta}\,|\,\boldsymbol{\eta} = \boldsymbol{0} \text{ on } \partial B_u\}$ – often called virtual displacement or test function. The following integration over the volume of the solid yields

$$\int_B \operatorname{Div}\boldsymbol{P}\cdot\boldsymbol{\eta}\,dV + \int_B \rho_0\,(\bar{\boldsymbol{b}} - \dot{\boldsymbol{v}})\cdot\boldsymbol{\eta}\,dV = 0\,. \tag{54}$$

By partial integration of the first term, application of the divergence theorem and introduction of the traction boundary condition one obtains the weak form of linear momentum

$$G(\boldsymbol{\varphi},\boldsymbol{\eta}) = \int_B \boldsymbol{P}\cdot\operatorname{Grad}\boldsymbol{\eta}\,dV - \int_B \rho_0\,(\bar{\boldsymbol{b}} - \dot{\boldsymbol{v}})\cdot\boldsymbol{\eta}\,dV - \int_{\partial B_\sigma} \bar{\boldsymbol{t}}\cdot\boldsymbol{\eta}\,dA = 0\,. \tag{55}$$

The gradient of the test function $\boldsymbol{\eta}$ can also be interpreted as variation $\delta\boldsymbol{F}$ of the deformation gradient. In the weak form (55) the first Piola-Kirchhoff stress tensor can be replaced through $\boldsymbol{P} = \boldsymbol{F}\,\boldsymbol{S}$ by the second

Piola-Kirchhoff stress tensor leading to

$$\boldsymbol{P}\cdot\mathrm{Grad}\,\boldsymbol{\eta} = \boldsymbol{S}\cdot\boldsymbol{F}^T\,\mathrm{Grad}\,\boldsymbol{\eta} = \boldsymbol{S}\cdot\frac{1}{2}\,(\boldsymbol{F}^T\,\mathrm{Grad}\,\boldsymbol{\eta}+\mathrm{Grad}^T\boldsymbol{\eta}\,\boldsymbol{F}) = \boldsymbol{S}\cdot\delta\boldsymbol{E}\,, \quad (56)$$

where the fact has been used that the scalar product of a symmetrical tensor (here $\boldsymbol{S}$) with a antisymmetrical part of a tensor is zero. $\delta\,\boldsymbol{E}$ denotes the variation of the Green-Lagrange strain tensor. Using (56) equation (55) can be rewritten as

$$G\,(\boldsymbol{\varphi}\,,\boldsymbol{\eta}\,) = \int\limits_B \boldsymbol{S}\cdot\delta\boldsymbol{E}\,dV - \int\limits_B \rho_0\,(\bar{\boldsymbol{b}}-\dot{\boldsymbol{v}})\cdot\boldsymbol{\eta}\,dV - \int\limits_{\partial B_\sigma} \bar{\boldsymbol{t}}\cdot\boldsymbol{\eta}\,dA = 0\,. \quad (57)$$

The first term in (57) denotes the internal virtual work also called stress divergence term. The last two terms describe the virtual work of the applied loading and the inertia term.

Weak form of linear momentum, current configuration The transformation of the weak form (55) to the current or spatial configuration is performed by kinematical operations in which the base vectors push forward to the configuration $\varphi(B)$. With the transformation of the first Piola-Kirchhoff stress tensor to the Cauchy stress tensor, see Table 3: $\boldsymbol{\sigma} = \frac{1}{J}\,\boldsymbol{P}\,\boldsymbol{F}^T$, the virtual work can be formulated as

$$\boldsymbol{P}\cdot\mathrm{Grad}\,\boldsymbol{\eta} = J\,\boldsymbol{\sigma}\,\boldsymbol{F}^{-T}\cdot\mathrm{Grad}\,\boldsymbol{\eta} = J\,\boldsymbol{\sigma}\cdot\mathrm{Grad}\,\boldsymbol{\eta}\,\boldsymbol{F}^{-1} = J\,\boldsymbol{\sigma}\cdot\mathrm{grad}\,\boldsymbol{\eta}\,.$$

Furthermore relations $dv = J\,dV \iff \rho = \rho_0\,J$ enable to write the weak form (55) in terms of the current configuration

$$g\,(\varphi,\boldsymbol{\eta}) = \int\limits_{\varphi(B)} \boldsymbol{\sigma}\cdot\mathrm{grad}\,\boldsymbol{\eta}\,dv - \int\limits_{\varphi(B)} \rho\,(\bar{\boldsymbol{b}}-\dot{\boldsymbol{v}})\cdot\boldsymbol{\eta}\,dv - \int\limits_{\varphi(\partial B_\sigma)} \hat{\boldsymbol{t}}\cdot\boldsymbol{\eta}\,da = 0\,. \quad (58)$$

In this relations equation (27) has been used to transform the traction vector $\bar{\boldsymbol{t}}$ to $\varphi(B)$. The symmetry of the Cauchy stress tensor facilitates the replacement of the spatial gradient of the test function $\boldsymbol{\eta}$ by its symmetric part. Hence with the definition

$$\nabla_x^S\,\boldsymbol{\eta} = \frac{1}{2}\,(\mathrm{grad}\,\boldsymbol{\eta} + \mathrm{grad}^T\,\boldsymbol{\eta}\,) \quad (59)$$

the weak form with respect to the spatial configuration follows

$$g\,(\varphi,\boldsymbol{\eta}\,) = \int\limits_{\varphi(B)} \boldsymbol{\sigma}\cdot\nabla_x^S\,\boldsymbol{\eta}\,dv - \int\limits_{\varphi(B)} \rho\,(\bar{\boldsymbol{b}}-\dot{\boldsymbol{v}})\cdot\boldsymbol{\eta}\,dv - \int\limits_{\varphi(\partial B_\sigma)} \hat{\boldsymbol{t}}\cdot\boldsymbol{\eta}\,da = 0\,. \quad (60)$$

This relation is, in a formal sense, equivalent to the principle of virtual work of the geometrically linear theory. But here the integral, the stress and virtual strain measures have to be evaluated with respect to the current configuration. Due to this the nonlinearities do appear, however hidden.

In the further variational formulations presented in this section we will neglect the inertia terms $\rho\,\dot{\boldsymbol{v}}$ and will concentrate on static equilibrium equations.

Weak form in the geometrically linear case For the geometrical linear theory the weak form is formally equivalent to (60). Only the operator defined in (59) has to be evaluated with respect to the coordinates of the initial configuration as well as the integrals

$$g^{lin}\,(\,\boldsymbol{u},\boldsymbol{\eta}\,) = \int\limits_{B} \boldsymbol{\sigma}\cdot\nabla^S_X\,\boldsymbol{\eta}\,dv - \int\limits_{B} \rho\,\bar{\boldsymbol{b}}\cdot\boldsymbol{\eta}\,dv - \int\limits_{\partial B_\sigma} \hat{\boldsymbol{t}}\cdot\boldsymbol{\eta}\,da = 0 \tag{61}$$

with $\nabla^S_X\,\boldsymbol{\eta} = \frac{1}{2}\,(\mathrm{Grad}\,\boldsymbol{\eta} + \mathrm{Grad}^T\,\boldsymbol{\eta}\,)$.

1.5 Variational functionals

Different variational functionals will be discussed which can alternatively be applied within the discretization process of the finite element method.

Principle of stationary elastic potential In case of a hyper elastic material there exist a strain energy function W, which describes the elastic energy stored in the solid. Based on this strain energy the classical principle of minimum of potential energy can be formulated in the geometrically linear theory. In finite deformation theory one has in general consider that deformations can occur which are unstable. Hence only a stationary value of the potential can be reached. We assume further that the applied loads are conservative which means path independent. For static problems one can state

$$\Pi\,(\boldsymbol{\varphi}) = \int\limits_{B} [\,W(\boldsymbol{C}) - \rho_0\,\bar{\boldsymbol{b}}\cdot\boldsymbol{\varphi}\,]\,dV - \int\limits_{\partial B_\sigma} \bar{\boldsymbol{t}}\cdot\boldsymbol{\varphi}\,dA \Longrightarrow STAT\,. \tag{62}$$

Of all possible deformations $\boldsymbol{\varphi}$ the ones which make Π stationary fulfil the equilibrium equation. The stationary value of (62) can be computed by the variation of Π with respect to the deformation. For this purpose we introduce the directional derivative

$$\delta\Pi = D\,\Pi\,(\boldsymbol{\varphi})\cdot\boldsymbol{\eta} = \frac{d}{d\alpha}\,\Pi\,(\boldsymbol{\varphi}+\alpha\,\boldsymbol{\eta})\Big|_{\alpha=0}\,, \tag{63}$$

which is also called first variation of Π. The application of this mathematical operation yields

$$D\,\Pi(\varphi)\cdot\boldsymbol{\eta} = \int_B [\,\frac{\partial W}{\partial \boldsymbol{C}}\cdot D\,\boldsymbol{C}\cdot\boldsymbol{\eta} - \rho_0\,\bar{\boldsymbol{b}}\cdot\boldsymbol{\eta}\,]\,dV - \int_{\partial B_\sigma} \bar{\boldsymbol{t}}\cdot\boldsymbol{\eta}\,dA = G(\boldsymbol{u},\boldsymbol{\eta}) = 0\,. \quad (64)$$

The directional derivative of the right Cauchy-Green strain tensor can easily be written in terms of the Green-Lagrange strain tensor:

$$\begin{aligned} D\boldsymbol{C}\cdot\boldsymbol{\eta} &= \frac{d}{d\alpha}\,\boldsymbol{F}^T\,(\varphi+\alpha\,\boldsymbol{\eta})\,\boldsymbol{F}\,(\varphi+\alpha\,\boldsymbol{\eta})\Big|_{\alpha=0} \\ &= \frac{d}{d\alpha}\,[\mathrm{Grad}\,(\varphi+\alpha\,\boldsymbol{\eta})]^T\;\mathrm{Grad}\,(\varphi+\alpha\,\boldsymbol{\eta})\Big|_{\alpha=0} \\ &= (\mathrm{Grad}\,\boldsymbol{\eta})^T\boldsymbol{F} + \boldsymbol{F}^T\,\mathrm{Grad}\,\boldsymbol{\eta} = 2\,D\,\boldsymbol{E}\cdot\boldsymbol{\eta} \end{aligned}$$

The last result corresponds to the variation $\delta\boldsymbol{E}$, defined already (56). The partial derivative of W with respect to $\boldsymbol{C}$ leads to the second Piola-Kirchhoff stress tensor $\boldsymbol{S}$, see (30): $\boldsymbol{S} = 2\,\partial W\,/\,\partial\boldsymbol{C}$. Hence equation (64) is equivalent to the weak form (57) for a hyperelastic material.

The construction of such principle has several advantages: (1) it can be the basis for mathematical investigations regarding existence and uniqueness of solutions, (2) it leads to the development of efficient algorithms for the solution of the nonlinear equations on the basis of optimization strategies.

Hu-Washizu principle for finite deformations Another variational principle is the Hu-Washizu principle, see Washizu (1975). It has gained significance during the last years for the construction of finite elements. This principles can be derived by writing the weak formulation with additional constraint equations which contain kinematics and constitutive equations. Due to this deformations, strains and stresses occur as independent variables. Contrary, once the principle is constructed its variation with respect to all variables yields the static equilibrium equations, the kinematical relations and the constitutive equation. The formulation of the nonlinear version of the Hu-Washizu principle can be given by any set of work conjugated variables. Here we will state it in terms of the deformation gradient $\boldsymbol{F}$, the first Piola-Kirchhoff stress tensor $\boldsymbol{P}$ and the Deformation φ

$$\begin{aligned} \Pi(\varphi,\boldsymbol{F},\boldsymbol{P}) &= \int_B [\,W(\boldsymbol{F}) + \boldsymbol{P}\cdot(\,\mathrm{Grad}\,\varphi - \boldsymbol{F}\,)\,]\,dV \\ &- \int_B \varphi\cdot\rho_0\,\bar{\boldsymbol{b}}\,dV - \int_{\partial B_\sigma} \varphi\cdot\hat{\boldsymbol{t}}\,dA \end{aligned} \quad (65)$$

The variation, according to the definition of the directional derivative given above, yields now three independent equations

$$
\begin{aligned}
D\Pi(\varphi, \boldsymbol{F}, \boldsymbol{P}) \cdot \boldsymbol{\eta} &= \int\limits_B (\, \boldsymbol{P} \cdot \operatorname{Grad} \boldsymbol{\eta} - \boldsymbol{\eta} \cdot \rho_0 \, \bar{\boldsymbol{b}}\,)\, dV - \int\limits_{\partial B_\sigma} \boldsymbol{\eta} \cdot \hat{\boldsymbol{t}}\, dA = 0\,, \\
D\Pi(\varphi, \boldsymbol{F}, \boldsymbol{P}) \cdot \boldsymbol{Q} &= \int\limits_B \boldsymbol{Q} \cdot (\, \operatorname{Grad} \varphi - \boldsymbol{F}\,)\, dV = 0\,, \qquad (66) \\
D\Pi(\varphi, \boldsymbol{F}, \boldsymbol{P}) \cdot \boldsymbol{A} &= \int\limits_B \boldsymbol{A} \cdot (\, \frac{\partial W}{\partial \boldsymbol{F}} - \boldsymbol{P}\,)\, dV = 0
\end{aligned}
$$

where $\boldsymbol{\eta}$, $\boldsymbol{Q}$ and $\boldsymbol{A}$ are the virtual displacements, the virtual stresses and the virtual strains, respectively. Observe that they represent the weak form (55), the kinematical relation (5) and a hyperelastic constitutive equation for $\boldsymbol{P}$. Note further that the last two terms can be interpreted as constraint terms which have to be fulfilled additionally together with the first equation. In that case $\boldsymbol{Q}$ and $\boldsymbol{A}$ are Lagrangian multipliers.

The variational principle (65) can also be written in terms of the second Piola-Kirchhoff stress and the Green-Lagrangian strains

$$
\begin{aligned}
\Pi(\boldsymbol{u}, \boldsymbol{E}, \boldsymbol{S}) &= \int\limits_B [\, \hat{W}(\boldsymbol{E}) + \boldsymbol{S} \cdot (\, \boldsymbol{E}(\boldsymbol{u}) - \boldsymbol{E}\,)\,]\, dV \\
&- \int\limits_B \boldsymbol{u} \cdot \rho_0 \, \bar{\boldsymbol{b}}\, dV - \int\limits_{\partial B_\sigma} \boldsymbol{u} \cdot \hat{\boldsymbol{t}}\, dA \qquad (67)
\end{aligned}
$$

where $\boldsymbol{E}(\boldsymbol{u}) = \frac{1}{2}\,(\, \boldsymbol{H} + \boldsymbol{H}^T + \boldsymbol{H}^T \boldsymbol{H}\,)$ with $\boldsymbol{H} = \operatorname{Grad} \boldsymbol{u}$, see Table 2. Its variation follow analogous to (66).

Hu-Washizu principle for small deformations The Hu-Washizu principle was originally derived in Washizu (1975) for small deformations. It then is stated in terms of the strain tensor $\boldsymbol{\epsilon}$, the first stress tensor $\boldsymbol{\sigma}$ and the displacement $\boldsymbol{u}$

$$
\begin{aligned}
\Pi(\boldsymbol{u}, \boldsymbol{\epsilon}, \boldsymbol{\sigma}) &= \int\limits_B [\, \frac{\Lambda}{2}\, (\operatorname{tr} \boldsymbol{\epsilon})^2 + \mu\, \boldsymbol{\epsilon} \cdot \boldsymbol{\epsilon} + \boldsymbol{\sigma} \cdot (\, \nabla_X^S \, \boldsymbol{u} - \boldsymbol{\epsilon}\,)\,]\, dV \\
&- \int\limits_B \varphi \cdot \rho_0 \, \bar{\boldsymbol{b}}\, dV - \int\limits_{\partial B_\sigma} \varphi \cdot \hat{\boldsymbol{t}}\, dA \qquad (68)
\end{aligned}
$$

The variation, according to the definition of the directional derivative given above, yields now three independent equations

$$
\begin{aligned}
D\Pi(\boldsymbol{u},\boldsymbol{\epsilon},\boldsymbol{\sigma})\cdot\boldsymbol{\eta} &= \int\limits_B (\boldsymbol{\sigma}\cdot\nabla_X^S\boldsymbol{\eta}-\boldsymbol{\eta}\cdot\rho_0\,\bar{\boldsymbol{b}})\,dV-\int\limits_{\partial B_\sigma}\boldsymbol{\eta}\cdot\hat{\boldsymbol{t}}\,dA=0\,,\\
D\Pi(\boldsymbol{u},\boldsymbol{\epsilon},\boldsymbol{\sigma})\cdot\boldsymbol{\tau} &= \int\limits_B \boldsymbol{\tau}\cdot(\nabla_X^S\boldsymbol{u}-\boldsymbol{\epsilon})\,dV=0\,, \qquad (69)\\
D\Pi(\boldsymbol{u},\boldsymbol{\epsilon},\boldsymbol{\sigma})\cdot\boldsymbol{\gamma} &= \int\limits_B \boldsymbol{\gamma}\cdot[\,(\Lambda\,\mathrm{tr}\boldsymbol{\epsilon}\,\mathbf{1}+2\,\mu\,\boldsymbol{\epsilon})-\boldsymbol{\sigma}\,]\,dV=0
\end{aligned}
$$

where $\boldsymbol{\eta}$, $\boldsymbol{\tau}$ and $\boldsymbol{\gamma}$ are the virtual displacements, the virtual stresses and the virtual strains, respectively. Observe that they represent the weak form (61), the kinematical relation are given in Table 2, and the linear elastic constitutive equation for $\boldsymbol{\sigma}$ follows from Table 4.

Hu-Washizu principle for incompressibility A special form of the Hu-Washizu variational principle can be applied for the construction of finite elements which have to represent nearly incompressible material behaviour. Since incompressibility is associated with a constraint for the volumetric deformation ($J = 1$), one can use the split (14) to distinguish volumetric and isochoric parts of the deformation. Based on this idea Simo et al. (1985) formulated a three-field functional which is only defined for the volumetric part of the deformation. Hence the independent variables are now the deformation φ, the pressure p and a strain variable θ which is equivalent to J. The last variable has to fulfil the constraint condition $\theta = J$. With the multiplicative split of the deformation gradient (14)

$$\bar{\boldsymbol{F}} = \theta^{\frac{1}{3}}\,\widehat{\boldsymbol{F}} \qquad (70)$$

one obtains in this case the split into volumetric and deviatoric parts. Note that we have in relation (70) $\widehat{\boldsymbol{F}} = J^{-\frac{1}{3}}\,\mathrm{Grad}\,\varphi$. Furthermore $\bar{\boldsymbol{C}} = \theta^{\frac{2}{3}}\,J^{-\frac{2}{3}}\,\boldsymbol{C} = \theta^{\frac{2}{3}}\,\widehat{\boldsymbol{C}}$ holds with (15). In this variational principle also the strain energy function $W(\boldsymbol{C})$, see (36), has to be defined on the basis of the new variables: $W(\boldsymbol{C}) = W(\theta^{\frac{2}{3}}\,\widehat{\boldsymbol{C}})$. Using the additive split $W = W(\theta) + W(\widehat{\boldsymbol{C}})$, see (36), the following three-field variational functional can be constructed

$$\begin{aligned}\Pi(\boldsymbol{\varphi},p,\theta) &= \int_B [\,W(\widehat{\boldsymbol{C}}) + W(\theta) + p(\,J-\theta\,)\,]\,dV \\ &- \int_B \boldsymbol{\varphi}\cdot\rho_0\,\bar{\boldsymbol{b}}\,dV - \int_{\partial B_\sigma} \boldsymbol{\varphi}\cdot\hat{\boldsymbol{t}}\,dA\,. \end{aligned} \tag{71}$$

By considering the relation (39) the Euler-Lagrange equations obtained from this variational principle are

$$\begin{aligned} D\Pi(\boldsymbol{\varphi},p,\theta)\cdot\boldsymbol{\eta} &= \int_B \{\,(\,\mathbb{P}[\,2\,\frac{\partial W}{\partial\widehat{\boldsymbol{C}}}] + p\,J\,\boldsymbol{C}^{-1}\,)\cdot\frac{1}{2}\,\delta\boldsymbol{C} - \boldsymbol{\eta}\cdot\rho_0\,\bar{\boldsymbol{b}}\,\}\,dV \\ &- \int_{\partial B_\sigma} \boldsymbol{\eta}\cdot\hat{\boldsymbol{t}}\,dA = 0 \\ D\Pi(\boldsymbol{\varphi},p,\theta)\,q &= \int_B q\,(\,J-\theta\,)\,dV = 0 \\ D\Pi(\boldsymbol{\varphi},p,\theta)\,\vartheta &= \int_B \vartheta\,(\frac{\partial W}{\partial\theta} - p\,)\,dV = 0\,, \end{aligned} \tag{72}$$

Here $\boldsymbol{\eta}$, q and ϑ are the virtual deformations, virtual pressure and virtual volumetric strains, respectively. By comparison of relation (40) with (41) one can conclude that one can use in equation $(72)_1$ also the expression $\boldsymbol{S}_{ISO} + \boldsymbol{S}_{VOL}$ for the first term in the integral. This explains the split into isochoric and volumetric parts.

The terms in equation $(72)_1$ are often written with respect to the spatial configuration since this simplifies the numerical implementation within the finite element method. With the variation of the Cauchy-Green tensor

$$\delta\,\boldsymbol{C} = \boldsymbol{F}^T\,\text{Grad}\,\boldsymbol{\eta} + \text{Grad}^T\,\boldsymbol{\eta}\,\boldsymbol{F} = \boldsymbol{F}^T\,(\,\text{grad}\,\boldsymbol{\eta} + \text{grad}^T\,\boldsymbol{\eta}\,)\,\boldsymbol{F} \tag{73}$$

one obtains with the definition (59)

$$\begin{aligned} D\Pi(\boldsymbol{\varphi},p,\theta)\cdot\boldsymbol{\eta} &= \int_B \{\,(\,\boldsymbol{F}\,\mathbb{P}[\,2\,\frac{\partial W}{\partial\widehat{\boldsymbol{C}}}]\,\boldsymbol{F}^T + p\,J\,\boldsymbol{1}\,)\cdot\nabla_x^S\,\boldsymbol{\eta} - \boldsymbol{\eta}\cdot\rho_0\,\bar{\boldsymbol{b}}\,\}\,dV \\ &- \int_{\partial B_\sigma} \boldsymbol{\eta}\cdot\hat{\boldsymbol{t}}\,dA = 0\,. \end{aligned} \tag{74}$$

The final result follows with (42) and (43)

$$
\begin{aligned}
D\Pi(\boldsymbol{\varphi}, p, \theta) \cdot \boldsymbol{\eta} &= \int\limits_B \{ (\boldsymbol{\tau}_{iso} \cdot \nabla_x^S \boldsymbol{\eta} + \tau_{vol} \operatorname{div} \boldsymbol{\eta} - \boldsymbol{\eta} \cdot \rho_0 \bar{\boldsymbol{b}} \} \, dV \\
&\quad - \int\limits_{\partial B_\sigma} \boldsymbol{\eta} \cdot \hat{\boldsymbol{t}} \, dA = 0 \qquad (75) \\
D\Pi(\boldsymbol{\varphi}, p, \theta) \, q &= \int\limits_B q \, (J - \theta) \, dV = 0 \qquad (76) \\
D\Pi(\boldsymbol{\varphi}, p, \theta) \, \vartheta &= \int\limits_B \vartheta \, (\frac{\partial W}{\partial \theta} - p) \, dV = 0 \, .
\end{aligned}
$$

In this formulations the integrands are given with respect to the spatial or current configurations whereas the integration still is performed in the initial configuration. The first equation denotes the weak form of the equilibrium (60) with the (Kirchhoff stresses instead of the Cauchy stresses). The second equation reproduces the constrain condition $J = \theta$ and the third equation yields the constitutive equation for the pressure p, see also $(43)_1$.

Hellinger-Reissner principle, small strains The Hu-Washizu principle is the most general principle which depends on three independent fields. Based on the fulfillment of the constraints we can eliminate one of the constraints directly. This yields e.g. a two-field formulation with displacements and stresses as primary variables. The adequate Hellinger-Reissner functional can be written for small strains as

$$
\Pi_{HR}(\boldsymbol{\sigma}\,, \boldsymbol{u}) = \int\limits_B \boldsymbol{\sigma} \cdot \nabla_X^S \boldsymbol{u} \, dV - \frac{1}{2} \int\limits_B \boldsymbol{\sigma} \cdot \mathbb{C}^{-1} [\boldsymbol{\sigma}] \, dV - \int\limits_B \hat{\boldsymbol{b}} \cdot \boldsymbol{u} \, dV - \int\limits_{\partial B_\sigma} \hat{\boldsymbol{t}} \cdot \boldsymbol{u} \, dA \, . \qquad (77)
$$

Here $\mathbb{C}$ is the constitutive tensor which relates strains and stresses in case of linear elasticity, seeTable 4,

$$
\boldsymbol{\sigma} = \mathbb{C}[\, \boldsymbol{\epsilon}(\boldsymbol{u}) \,] = \mathbb{C}[\, \nabla_X^S \boldsymbol{u} \,] \, . \qquad (78)
$$

$\hat{\boldsymbol{b}}$ and $\hat{\boldsymbol{t}}$ are the applied body forces and boundary tractions, respectively.

The variation of this variational functional yields

$$
\begin{aligned}
D\Pi \cdot \boldsymbol{\eta} &= \int\limits_B \boldsymbol{\sigma} \cdot \nabla_X^S \boldsymbol{\eta} \, dV - \int\limits_B \hat{\boldsymbol{b}} \cdot \boldsymbol{\eta} \, dV - \int\limits_{\partial B_\sigma} \hat{\boldsymbol{t}} \cdot \boldsymbol{\eta} \, dA = 0 \, , \\
D\Pi \cdot \boldsymbol{\tau} &= \int\limits_B \boldsymbol{\tau} \cdot \nabla^S \boldsymbol{u} \, dV - \int\limits_B \boldsymbol{\tau} \cdot \mathbb{C}^{-1} \, [\boldsymbol{\sigma}] \, dA = 0 \qquad (79)
\end{aligned}
$$

where $\boldsymbol{\tau}$ and $\boldsymbol{\eta}$ are the variations of the stresses and the displacements, respectively. The first equation represents the weak form of equilibrium. The second equation is the weak statement of the constitutive equation. It could also be written as

$$\int_B \boldsymbol{\tau} \cdot (\nabla_X^S \boldsymbol{u} - \mathbb{C}^{-1}[\boldsymbol{\sigma}])\, dV = 0 \tag{80}$$

since the last term is nothing else then the inversion of (78). Hence the two-field Hellinger-Reissner principle has besides the weak form of equilibrium one additional constraint equation, which is the inverse of constitutive equation (78).

More general, the relation stemming from the Legendre transformation

$$W(\boldsymbol{\epsilon}) = \boldsymbol{\sigma} \cdot \boldsymbol{\epsilon} - U(\boldsymbol{\sigma}) \tag{81}$$

is valid. In linear elasticity the strain energy has the form $W(\boldsymbol{\epsilon}) = \frac{1}{2}\boldsymbol{\varepsilon} \cdot \mathbb{C}[\boldsymbol{\epsilon}]$ which yields with (78)

$$U(\boldsymbol{\sigma}) = \frac{1}{2}\boldsymbol{\sigma} \cdot \mathbb{C}^{-1}[\boldsymbol{\sigma}]. \tag{82}$$

Hence it can be seen that the first two terms in (77) are equivalent to the integral of the strain energy $W(\boldsymbol{\epsilon})$.

Dual mixed formulation Another version has as only unknowns the stresses. This is based on the dual formulation

$$U(\boldsymbol{\sigma}) = \frac{1}{2}\int_B \boldsymbol{\sigma} \cdot \mathbb{C}^{-1}[\boldsymbol{\sigma}]\, dV \tag{83}$$

in which the complementary energy $U(\boldsymbol{\sigma})$ has to be minimized by a stress field which fulfills the strong form of linear and angular momentum (23) and (26) in B for the case of statics and the the stress boundary conditions (Neumann boundary conditions) on ∂B_σ. It is of course complicated to find a stress field which fulfills these local conditions. Hence one can add the linear momentum to (83) by using the Lagrangian multiplier method

$$\frac{1}{2}\int_B \boldsymbol{\sigma} \cdot \mathbb{C}^{-1}[\boldsymbol{\sigma}]\, dV + \int_B (\operatorname{div}\boldsymbol{\sigma} + \rho\, \boldsymbol{b}) \cdot \boldsymbol{u}\, dV \rightarrow EXTR \tag{84}$$

Now the stresses have to be computed for fields which fulfill the divergence operator and the angular momentum (26) in B and the Neumann boundary conditions on ∂B_σ. The Lagrange multiplier $\boldsymbol{u}$ can be interpreted as a displacement. Since it is not simple to find stable finite element discretizations for (84), see Brezzi and Fortin (1991), one can relax the strong symmetry condition (26) of the stress field , see e.g. Fraeijs de Veubeke (1975), Arnold et al. (1984) and Stenberg (1988). This yields the modified functional

$$\frac{1}{2}\int_B \boldsymbol{\sigma}\cdot\mathbb{C}^{-1}\,[\boldsymbol{\sigma}]\,dV + \int_B (\,\mathrm{div}\boldsymbol{\sigma} + \rho\,\boldsymbol{b}\,)\cdot\boldsymbol{u}\,dV + \int_B \boldsymbol{\sigma}\cdot\boldsymbol{\beta}\,dV \rightarrow EXTR \tag{85}$$

Here the stress field has to fulfill the the divergence operator and the Neumann boundary conditions. The additional Lagrange multiplier $\boldsymbol{\gamma}$ is associated with the linear rotation field. At the extremum it has the value $\boldsymbol{\beta} = \frac{1}{2}\,(\,\nabla\boldsymbol{u} - \nabla^T\boldsymbol{u}\,)$. Variation of (85) leads to the weak form for the three independent fields $\boldsymbol{\sigma}$, $\boldsymbol{u}$ and $\boldsymbol{\beta}$

$$\begin{aligned}\int_B \boldsymbol{\tau}\cdot\mathbb{C}^{-1}\,[\boldsymbol{\sigma}]\,dV + \int_B \mathrm{div}\boldsymbol{\tau}\cdot\boldsymbol{u} + \int_B \boldsymbol{\tau}\cdot\boldsymbol{\beta}\,dV &= 0\\ \int_B (\,\mathrm{div}\boldsymbol{\sigma} + \rho\,\boldsymbol{b}\,)\cdot\boldsymbol{\eta}\,dV &= 0\\ \int_B \boldsymbol{\sigma}\cdot\boldsymbol{\gamma}\,dV &= 0\end{aligned} \tag{86}$$

Here $\boldsymbol{\tau}$, $\boldsymbol{\eta}$ and $\boldsymbol{\gamma}$ are the test functions. This set of equations can also be obtained from the strong forms. Then one has to multiply linear and angular momentum by $\boldsymbol{\eta}$ and $\boldsymbol{\gamma}$, respectively, which yields $(86)_{2,3}$. Equation $(86)_1$ follows by multiplying the linear constitutive equation (78), $\mathbb{C}^{-1}\,[\boldsymbol{\sigma}] - \boldsymbol{\epsilon}(\boldsymbol{u}) = \boldsymbol{0}$ by a test function $\boldsymbol{\tau}$ which is from the stress space fulfilling the divergence operator. Additional partial integration yields the final result.

Hellinger-Reissner principle, finite strains For finite elastic deformations we consider the Neo- Hooke constitutive equation

$$\boldsymbol{S} = \frac{\Lambda}{2}\,(J^2 - 1)\,\boldsymbol{C}^{-1} + \mu\,(\,\boldsymbol{1} - \boldsymbol{C}^{-1}\,) \tag{87}$$

and assume that it is possible to invert this equation

$$\boldsymbol{C} = \boldsymbol{f}\,(\boldsymbol{S})\,. \tag{88}$$

Since the Legendre transformation cannot be applied in the finite deformation case we formulate the principle analogous to equation (79). This yields

$$
\begin{aligned}
D\Pi \cdot \boldsymbol{\eta} &= \int_B \frac{1}{2}\, \boldsymbol{S} \cdot \boldsymbol{C}(\boldsymbol{\eta})\, dV - \int_B \hat{\boldsymbol{b}} \cdot \boldsymbol{\eta}\, dV - \int_{\partial B_\sigma} \hat{\boldsymbol{t}} \cdot \boldsymbol{\eta}\, dA = 0\,, \\
D\Pi \cdot \boldsymbol{Q} &= \int_B \frac{1}{2}\, \boldsymbol{Q} \cdot [\, \boldsymbol{C}(\boldsymbol{u}) - \boldsymbol{f}\,(\boldsymbol{S})\,]\, dA = 0\,. \qquad (89)
\end{aligned}
$$

where $\boldsymbol{Q}$ is the virtual stress, $\boldsymbol{C}(\boldsymbol{\eta}) = \boldsymbol{F}^T\, \text{Grad}\, \boldsymbol{\eta} + \text{Grad}^T\, \boldsymbol{\eta}\, \boldsymbol{F}$ is the virtual strain and $\boldsymbol{C}(\boldsymbol{u}) = \boldsymbol{F}^T\, \boldsymbol{F}$ is the right Cauchy-Green tensor, the latter depending only on the deformation $\boldsymbol{\varphi}$.

Note that the inverse (88) is not uniquely defined, either locally or globally, see Ogden (1984). However we can invert (87) by looking at different solution branches. For this we rewrite (87) as

$$
\hat{\boldsymbol{A}}(\boldsymbol{S}) = \hat{\beta}(\boldsymbol{C})\, \boldsymbol{C}^{-1} \qquad (90)
$$

with

$$
\hat{\beta} = \alpha - J^2\,, \quad \alpha = 1 + \frac{2\mu}{\Lambda}\,, \quad \hat{\boldsymbol{A}}(\boldsymbol{S}) = \frac{2}{\Lambda}\,(\, \mu \boldsymbol{1} - \boldsymbol{S}\,) \qquad (91)
$$

The multiplication of (90) with $\hat{\boldsymbol{A}}^{-1}$ from the left side leads to

$$
\boldsymbol{C} = \hat{\beta}(\boldsymbol{C})\, \hat{\boldsymbol{A}}^{-1}(\boldsymbol{S}) \qquad (92)
$$

Now it remains to compute $\hat{\beta}(\boldsymbol{C})$ in dependence of $\hat{\boldsymbol{A}}$. The computation of the determinant of (92) yields

$$
\det \hat{\boldsymbol{A}} = \frac{\hat{\beta}^3}{J^2} \qquad (93)
$$

since $J^2 = \det \boldsymbol{C}$. With the abbreviation $\hat{a} = \det \hat{\boldsymbol{A}}$ and $(91)_1$ this leads to a cubic equation for $\hat{\beta}$:

$$
\hat{\beta}^3 + \hat{\alpha}\, \hat{\beta} - \hat{a}\alpha = 0\,. \qquad (94)
$$

To obtain a simpler solution one can define $\beta = c_1\, \hat{\beta}$, $\boldsymbol{A} = c_2\, \hat{\boldsymbol{A}}$ and $a = \det \boldsymbol{A}$. Together with c_1 and c_2

$$
c_1 = \frac{2}{3\,\alpha}\,, \qquad c_2 = \frac{1}{3} \left[\frac{4}{\alpha^2} \right]^{\frac{1}{3}}\,, \qquad (95)
$$

which only depend on the Lame constants, this provides the cubic equation for β

$$
\beta^3 + 3\, a\, \beta - 2\, a = 0\,. \qquad (96)
$$

Depending on the discriminant $D = a^3 + a^2$ we have to distinguish three different solutions:

- $D > 0$: For this case there (96) has only one real solution

$$\beta = r - \frac{a}{r} \qquad \text{with } r = \left[a + \sqrt{a^3 + a^2} \right]^{\frac{1}{3}} . \tag{97}$$

- $D < 0$: This case is equivalent to $a < -1$ and yields three solutions for (96)

$$\beta = -2\sqrt{-a}\,\cos\left[\frac{1}{3}\left(\arccos\frac{1}{\sqrt{-a}} + 2\pi k\right)\right], \qquad k = 0, 1, 2 \tag{98}$$

  In the physical problem we have to fulfil $J > 0$. From $(91)_1$ and $(95)_1$ it then follows that $\beta < \frac{2}{3}$. Hence only the solution with $k = 0$ remains under these circumstances

$$\beta = -2\sqrt{-a}\,\cos\left(\frac{1}{3}\arccos\frac{1}{\sqrt{-a}}\right) \tag{99}$$

- $D = 0$: In this case the determinant a either is $a = 0$ or $a = -1$. For $a = 0$ the only solution is $\beta = 0$ which yields with $(91)_1$ and $(95)_1$ for the Jacobian $J = \sqrt{\alpha}$. For $a = -1$ β can be obtained from (99). This yields in the limit $\beta = -2$.

With this solution we can now, based on $(91)_1$ and $(91)_3$, derive the expression for the inverse of the Neo-Hooke material (87). With

$$\boldsymbol{A} = \frac{2}{3\,\Lambda}\left[\frac{4}{\alpha^2}\right]^{\frac{1}{3}} (\mu\mathbf{1} - \boldsymbol{S}) \tag{100}$$

we obtain the final result

$$\boldsymbol{C} = \beta\left[\frac{\alpha}{2}\right]^{\frac{1}{3}} \boldsymbol{A}^{-1} = \frac{3}{4}\,\beta\,\alpha\,\Lambda\,(\mu\mathbf{1} - \boldsymbol{S})^{-1} \tag{101}$$

1.6 Linearization of variational formulations

For the solution of the set of nonlinear equations many different algorithms are known. Often Newton's method is applied since it possesses the advantage of a quadratic convergence close to the solution point. In case of Newton's method, an improved solution is obtained from the Taylor series expansion of the nonlinear equation at the already computed approximate solution. This Taylor expansion corresponds in finite element applications to the linearization of the weak form, or in solid mechanics to the linearization of the virtual work or variational principles and can be obtained by the directional derivative.

The linearization will be stated first for the weak form with reference to the initial configuration (55). In general we will compute the linearization at a deformation state of the solid which is in equilibrium. This state will be denoted by $\bar{\varphi}$.

The linear part of the weak form is given by

$$L\,[\,G\,]_{\varphi=\bar{\varphi}} = G\,(\bar{\varphi},\boldsymbol{\eta}) + DG\,(\bar{\varphi},\boldsymbol{\eta}) \cdot \Delta\boldsymbol{u}\,. \tag{102}$$

The operator $G(\bar{\varphi},\boldsymbol{\eta})$ corresponds to (56), only the state $\varphi = \bar{\varphi}$ is inserted at which the linearization will be computed. By assuming that the load is conservative one can compute the directional derivative of G in the direction of $\Delta\boldsymbol{u}$ from the first term in (56)

$$D\,G\,(\bar{\varphi},\boldsymbol{\eta}) \cdot \Delta\boldsymbol{u} = \int\limits_B [D\boldsymbol{P}(\bar{\varphi}) \cdot \Delta\boldsymbol{u}] \cdot \text{Grad}\,\boldsymbol{\eta}\,dV\,, \tag{103}$$

hence all other terms are independent of the deformation. The linearization of the first Piola-Kirchhoff stress tensor yields with $\boldsymbol{P} = \boldsymbol{F}\,\boldsymbol{S}$

$$DG\,(\bar{\varphi},\boldsymbol{\eta}) \cdot \Delta\boldsymbol{u} = \int\limits_B \{\,\text{Grad}\,\Delta\boldsymbol{u}\,\bar{\boldsymbol{S}} + \bar{\boldsymbol{F}}\,[D\boldsymbol{S}(\bar{\varphi}) \cdot \Delta\boldsymbol{u}]\,\} \cdot \text{Grad}\,\boldsymbol{\eta}\,dV\,. \tag{104}$$

Terms with a bar have to be evaluated at the deformation state $\bar{\varphi}$. The linearization of the second Piola-Kirchhoff stress tensor follows as

$$D\boldsymbol{S}(\bar{\varphi}) \cdot \Delta\boldsymbol{u} = \bar{\mathbb{C}}\,[\,\Delta\bar{\boldsymbol{E}}\,]\,, \tag{105}$$

where the last term is the linearization of the Green-Lagrange strain tensor $\boldsymbol{E}$ at $\bar{\varphi}$. The elasticity tensor $\mathbb{C}_R$ which is also referred to the initial configuration B is given by

$$\bar{\mathbb{C}} = 4\,\frac{\partial^2 W}{\partial\boldsymbol{C}\,\partial\boldsymbol{C}}\bigg|_{\varphi=\bar{\varphi}} \tag{106}$$

at the state $\bar{\varphi}$.

The use of (106) in (104) completes the linearization

$$DG(\bar{\varphi},\boldsymbol{\eta}) \cdot \Delta\boldsymbol{u} = \int\limits_B \{\,\text{Grad}\,\Delta\boldsymbol{u}\,\bar{\boldsymbol{S}} + \bar{\boldsymbol{F}}\,\bar{\mathbb{C}}\,[\Delta\bar{\boldsymbol{E}}\,]\,\} \cdot \text{Grad}\,\boldsymbol{\eta}\,dV\,. \tag{107}$$

Note that also $\bar{\mathbb{C}}$ has to be evaluated at state $\bar{\varphi}$. By applying the trace operation to the second term and by using the symmetry of $\bar{\mathbb{C}}$ one can find a more compact form of (107)

$$DG(\bar{\varphi},\boldsymbol{\eta}) \cdot \Delta\boldsymbol{u} = \int\limits_B \{\,\text{Grad}\,\Delta\boldsymbol{u}\,\bar{\boldsymbol{S}} \cdot \text{Grad}\,\boldsymbol{\eta} + \delta\bar{\boldsymbol{E}} \cdot \bar{\mathbb{C}}\,[\Delta\bar{\boldsymbol{E}}\,]\,\}\,dV\,. \tag{108}$$

Here the symmetry of the linearization operation is $\boldsymbol{\eta}$ and $\Delta \boldsymbol{u}$ can be observed. The first term in equation (108) is often named initial stress term since the stresses at the given state appear directly. The second term contains besides the incremental constitutive tensor $\bar{\mathbb{C}}$, also the variation of the Green-Lagrange strain tensor $\delta \bar{\boldsymbol{E}} = \frac{1}{2}(\bar{\boldsymbol{F}}^T \operatorname{Grad} \boldsymbol{\eta} + \operatorname{Grad}^T \boldsymbol{\eta}\, \bar{\boldsymbol{F}})$ and the increment of the Green-Lagrange strain tensor $\Delta \bar{\boldsymbol{E}} = \frac{1}{2}(\bar{\boldsymbol{F}}^T \operatorname{Grad} \Delta \boldsymbol{u} + \operatorname{Grad}^T \Delta \boldsymbol{u}\, \bar{\boldsymbol{F}})$.

The linearization of the principle of virtual work in terms of the current configuration is obtained by a push forward of the linearization (108) to the already computed configuration $\bar{\varphi}$. Using the transformations for the linearization of the Green-Lagrange strain tensor which resulted as push forward in $\nabla^S_{\bar{x}} \Delta \boldsymbol{u}$, it follows for the second term in (108)

$$\int_B \nabla^S_{\bar{x}} \boldsymbol{\eta} \cdot \bar{\mathbb{D}} \, [\nabla^S_{\bar{x}} \Delta \boldsymbol{u}] \, dV \, .$$

In this equations one can compute the fourth order tensor $\bar{\mathbb{D}}$ from $\bar{\mathbb{C}}$, see e.g. Wriggers (2008).

The first term in (108) can be directly recast with $\bar{\boldsymbol{\tau}} = \bar{\boldsymbol{F}}\, \bar{\boldsymbol{S}}\, \bar{\boldsymbol{F}}^T$ as

$$\operatorname{Grad} \Delta \boldsymbol{u}\, \bar{\boldsymbol{S}} \cdot \operatorname{Grad} \boldsymbol{\eta} = \bar{\boldsymbol{F}} \operatorname{Grad} \Delta \boldsymbol{u}\, \bar{\boldsymbol{F}}^{-1}\, \bar{\boldsymbol{\tau}}\, \bar{\boldsymbol{F}}^{-1} \cdot \operatorname{Grad} \boldsymbol{\eta} = \nabla_{\bar{x}} \Delta \boldsymbol{u}\, \bar{\boldsymbol{\tau}} \cdot \nabla_{\bar{x}} \boldsymbol{\eta} \, . \tag{109}$$

This results in the linearization with respect to the known current configuration $\bar{\varphi}$ where all quantities have to be evaluated at $\bar{\varphi}$.

$$Dg(\bar{\varphi}, \boldsymbol{\eta}) \cdot \Delta \boldsymbol{u} = \int_B \{ \nabla_{\bar{x}} \Delta \boldsymbol{u}\, \bar{\boldsymbol{\tau}} \cdot \nabla_{\bar{x}} \boldsymbol{\eta} + \nabla^S_{\bar{x}} \boldsymbol{\eta} \cdot \bar{\mathbb{D}} \, [\nabla^S_{\bar{x}} \Delta \boldsymbol{u}] \} \, dV \, . \tag{110}$$

The integral (110) can now be referred to the current configuration with the relation $d\bar{v} = \bar{J} dV$. For this purpose we introduce Cauchy stress tensor $\bar{\boldsymbol{\sigma}} = \frac{1}{\bar{J}} \bar{\boldsymbol{\tau}}$ and define a further incremental constitutive tensor

$$\hat{\bar{\mathbb{D}}} = \frac{1}{\bar{J}} \bar{\mathbb{D}} \tag{111}$$

such that we obtain

$$Dg(\bar{\varphi}, \boldsymbol{\eta}) \cdot \Delta \boldsymbol{u} = \int_{\bar{\varphi}(B)} \{ \nabla_{\bar{x}} \Delta \boldsymbol{u}\, \bar{\boldsymbol{\sigma}} \cdot \nabla_{\bar{x}} \boldsymbol{\eta} + \nabla^S_{\bar{x}} \boldsymbol{\eta} \cdot \hat{\bar{\mathbb{D}}} \, [\nabla^S_{\bar{x}} \Delta \boldsymbol{u}] \} \, dv \, . \tag{112}$$

This relation is known in the literature also as updated Lagrange formulation, see e.g. Bathe et al. (1975) or Bathe (1996), hence the deformation state $\bar{\varphi}$ to which the formulation is referred is only obtained within the nonlinear solution process. The state $\bar{\varphi}$ is computed within an incremental solution procedure by an updating of all deformation states in a successive manner.

With the above given linearizations all relations needed within the Newton method are available as well as for formulations with respect to the initial configuration as with respect to the current configuration. These relations can be used in the same way for the linearizations of other variational principles discussed in the previous sections [5].

2 Mixed Finite Element Methods

2.1 Requirements for continuum finite elements

The search for well performing finite elements which can be applied to arbitrary problem classes within solid mechanics has a long history, as one can see from the numerous scientific papers devoted to this topic. Main targets of a development of finite elements are:

- Locking free behaviour for incompressible material,
- good bending performance,
- no locking in thin elements,
- no sensitivity against mesh distortions,
- good coarse mesh accuracy,
- simple implementation of nonlinear constitutive equations and
- efficiency (e.g. few necessary integration points).

The first point is associated with the numerical simulation of special problem classes which include in solid mechanics rubber like materials and elasto-plastic material equations in the framework of J_2-plasticity. The constraint related to the incompressible behaviour leads even for geometrical linear elements to locking, see e.g. Braess (1992) and Hughes (1987).

[5] It should be emphasized that the linearization of the continuous problem as stated in (56) does not coincide in all cases with the linearization of the discrete problem which results from a finite element discretization. In case of a finite element formulation of a static problem with continuous interpolations functions for the displacements one can show the equivalence of the discretization of the linearizations given above and the direct linearization of the discretization of the weak form using finite elements. In case that the finite element method is applied to problems of elasto-plasticity or other inelastic constitutive equations then the linearization of the discrete form is no longer equivalent to the discretization of the so far given linearizations since the integration algorithm for the evolution equations of the inelastic material response plays a prominent role and hence the linearization depends on this algorithm, see e.g. Simo et al. (1985) or for a more general overview Simo and Hughes (1998).

The second and third points are of significance when three-dimensional solid elements shall be employed to solve beam- or shell problems since beam and shell structures are often dominated by bending behaviour and are by construction thin in one or two spatial coordinates.

The fourth point is essential when modern methods for mesh generation are employed. These methods lead for arbitrary geometries to so called unstructured meshes which consist of finite elements shapes with arbitrary geometry.

The fifth point is related to the fact that one hasoften to analyze three-dimensional components in real engineering applications which size and complexity cannot be modelled using a converged mesh, especially when the simulation is nonlinear. Hence there is still need for elements which depict a good accuracy, even when used within a coarse mesh.

Engineering applications demand often the use of new nonlinear constitutive equations. Here a simple interface to the finite element should support the user in order to efficiently change existing finite elements and to be able to implement complex constitutive equations.

Finally we like to mention that efficiency is not only related to speed of the element formulation but also to the memory requirement. The latter demand is essential when e.g. inelastic problems with several hundred thousand or millions of finite element have to solved within a given time frame. Speed of the element formulation is essential when iterative solvers are applied since in that case the time for the computation of residuals and tangent matrices is of the same order as the time used by the solver within one iteration.

It is well known that a pure displacement element with bi- or tri-linear ansatz functions has bad convergence behaviour in bending problems and within applications with incompressible materials. Thus different variational formulations have been explored in order to construct finite elements which fulfill as many requirements stated above as possible. Among these are

- reduced integration,
- stabilization methods,
- hybrid or mixed variational principles which base on complimentary energy written in terms of the stress field,
- mixed variational principle of Hu-Washizu type and
- mixed variational principle for special quantities.

In the following we will summarize different possibilities and discuss their differences. Then some of the techniques are presented in detail.

1. **Reduced integration and stabilization.** The most simple method is the "underintegration" of the integrals leading to the finite element

vectors and matrices. It is also very efficient and safes memory for history data storage since less integration points are used. Underintegration or reduced integration means that less Gauss points are used for the integration of tangent matrices and residual vectors than necessary for the chosen polynomial degree of the shape functions, for first applications see e.g. Zienkiewicz et al. (1971). Reduced integration was developed to avoid locking in case of incompressibility. In that case it is often only applied to the pressure part of the constitutive equation, see e.g. Malkus and Hughes (1978) and Hughes (1980). Reduced integration is always associated with a rank deficiency of the tangent matrices which can be cured by different methods, generally known as stabilization techniques, see e.g. Belytschko et al. (1984) and Reese and Wriggers (2000). Using the reduced integration together with stabilization leads to finite elements which fulfill conditions 1., 4., 5., 6. and 7. Hence these elements are locking free in case of incompressibility, they have a good coarse mesh accuracy, are not sensitive against mesh distortions and can be used for arbitrary constitutive equations. The reduced integration provides the most efficient possibility to computed the element residual and the element tangent stiffness (e.g. for an eight-node brick element only one Gauss point is needed). However these elements need often the choice of artificial stabilization parameters. In the worst case, e.g. for some bending problems, the finite element solution can directly depend on the stabilization parameter.

2. **Hybrid or mixed variational principles.** When mixed variational principles are used as basis for finite element diecsretization then different possibilities exist for the construction of the finite element matrices. This is due to the fact that one can state many different mixed forms, see last section. Some of them need conforming displacement fields together with non-conforming stress or strain fields, others rely on conforming stress fields but allow non-conforming displacement fields. For the case of linear elasticity hybrid elements where first described in Pian (1964) which has lead to many different finite element formulations up to now. Within this approach Pian and Sumihara (1984) developed a finite element which is efficient and accurate. However due to the need to invert the constitutive equations within the formulation in order to obtain the constitutive equations in terms of the stress field, there are only few elements for St. Venant materials known which work for large deformations. A special formulation for Neo-Hooke materials will be presented in section 2.3.
3. **Enhanced strain elements based on the Hu-Washizu prin-**

ciple. Within the enhanced strain formulations, see Simo and Rifai (1990), non-conforming strain measures are introduced within the Hu-Washizu principle. In follow up work, Simo and Armero (1992) and Simo et al. (1993) have derived a family of enhanced elements for large deformations and inelastic constitutive equations based in the Hu-Washizu. This class of elements is related to the incompatible mode elements which were developed by Wilson et al. (1973) and Taylor et al. (1976) for linear problems. The enhanced strain elements fulfill point 1. to 6. of the above mentioned requirements. Hence they are well suited for all applications. However these elements have some disadvantages. They need a statical condensation on element level which reduces their efficiency. Furthermore the degrees of freedom belonging to the enhanced strains need additional storage on element level. A further point is related to the hour-glassing of the enhanced strain elements under pressure. This fact was discovered by Wriggers and Reese (1996). Solutions which partly solve this problem are provided in Korelc and Wriggers (1996), Glaser and Armero (1997), Reese and Wriggers (2000) and Reese (2005) where different methods have been used to overcome the hour glassing.

4. **Mixed variational principles for problems with rotational degrees of freedom.** When not only the the momentum is weakly enforced, but also the moment of momentum, which usually leads, see (26), to the symmetry of the stress tensor, then rotational degrees of freedom can be introduced as independent field variables. Finite elements which base on such formulation were constructed in e.g. Hughes and Brezzi (1989). Further applications of such variational formulations can be found for two-dimensional elements which are a basis for shell formulations, see e.g. Ibrahimbegovic et al. (1990), Iura and Atluri (1992) and Gruttmann et al. (1992).
5. **Mixed variational principles for special quantities.** Often problems have to be considered which include special constraint conditions. In such cases it is advantageous to formulate mixed principles which are tailored to fulfill such constraint conditions. Examples are solid elements for plates or shells where for thin structures the transverse shear becomes zero in the limit. Another example is the constraint related to incompressibility. This constraint occurs in rubber elasticity and in case of plastic J_2-flow. The related special variational principle relies on a split of the kinematical variables into volumetric and deviatoric parts.

   The related finite elements fulfill point 1., 4., 5. and 7. of the above mentioned requirements. The formulation of the constitutive eqations is often more complex than in the standard formulation. This is especially true for large deformations making the linearizations, needed within the Newton method, more complex.

2.2 Mixed elements for incompressibility

Pure displacement elements are not suitable for problems in which the constitutive behaviour exhibit incompressibility since they tend to locking. Locking means in this connection that the constraint conditions due to incompressibility which are related to the pure volumetric mode (in the elastic case we have $J = \det \boldsymbol{F} = 1$ and for plastic flow the condition $J_p = \det \boldsymbol{F}_p = 1$ holds) can only be fulfilled with an considerable stiffening of the bending modes. Thus this behaviour is also called volume locking. Mixed finite element methods can help to avoid this type of locking, see e.g. Zienkiewicz and Taylor (1989) and Brezzi and Fortin (1991). There exist different possibilities to construct mixed elements. These are discussed in the following.

- **Method of Lagrangian multipliers.** Here the constraint condition of incompressibility will be directly introduced via the methods of Lagrangian multipliers. For this one formulates the strain energy

$$W = W_{inkomp} + p\,G(J) \quad \text{mit} \quad G(J) = 0\,. \tag{113}$$

The constraint condition is then given for finite deformations as $G(J) = J - 1$ with $J =$ is e.g. given by the Mooney-Rivlin material, see Table 4. Finite elements which are based on this methodology have the disadvantage that contrary to the pure displacement elements additional unknowns occur. These are the Lagrangian multipliers which are equivalent to the pressure p. Hence special techniques are needed to solve the associated incremental equation system for displacements and Lagrangian multipliers

$$\begin{bmatrix} \boldsymbol{K}_{T\,uu} & \boldsymbol{B}_{T\,up} \\ \boldsymbol{B}_{T\,pu}^T & \boldsymbol{0} \end{bmatrix} \begin{Bmatrix} \Delta \boldsymbol{u} \\ \Delta \boldsymbol{p} \end{Bmatrix} = - \begin{Bmatrix} \boldsymbol{R}_u \\ \boldsymbol{R}_p \end{Bmatrix} \tag{114}$$

which has zero entries in the diagonal[6]. The sub-matrix $\boldsymbol{K}_{T\,uu}$ follows from W_{inkomp} while $\boldsymbol{B}_{T\,up}$ is related to the discretization of the term $p\,G(J)$. Associated finite element formulations can be found in Oden and Key (1970) and Duffet and Reddy (1983).

[6]Modern sparse equation solvers like PARDISO (Schenk and Gärtner (2004)) solve such systems in an efficient way.

- **Perturbed Lagrangian method.** To have a greater variability for the formulation of ansatz functions one can introduce the following strain energy function

$$W = W_{inkomp} + p\,G(J) - \frac{1}{2\,\epsilon}\,p^2\,. \tag{115}$$

The constraint condition is again given by $G(J) = J - 1$. $\epsilon > 0$ is a perturbation parameter. Choosing now continuous ansatz function for displacements and pressure the following incremental equation system can be derived

$$\begin{bmatrix} \boldsymbol{K}_{T\,uu} & \boldsymbol{B}_{T\,up} \\ \boldsymbol{B}_{T\,pu}^{T} & -\frac{1}{\epsilon}\,\boldsymbol{K}_{pp} \end{bmatrix} \begin{Bmatrix} \Delta \boldsymbol{u} \\ \Delta \boldsymbol{p} \end{Bmatrix} = - \begin{Bmatrix} \boldsymbol{R}_u \\ \boldsymbol{R}_{p\,\epsilon} \end{Bmatrix}\,. \tag{116}$$

Here contrary to (114) the incremental displacements and pressures can be computed using standard equation solvers. The pressures can be removed from the system by using the Schur complement. This leads to

$$\left[\boldsymbol{K}_{T\,uu} + \epsilon\,\boldsymbol{B}_{T\,up}\,\boldsymbol{K}_{pp}^{-1}\,\boldsymbol{B}_{T\,pu}^{T} \right] \Delta \boldsymbol{u} = -\boldsymbol{R}_u - \epsilon\,\boldsymbol{B}_{T\,up}\,\boldsymbol{K}_{pp}^{-1}\,\boldsymbol{R}_{p,\epsilon}\,. \tag{117}$$

When discontinuous ansatz functions are used for the pressure variables then the pressures can be eliminated on element level[7]. Note that the solution now depends on the perturbation parameter. For small values of ϵ the influence of the constraint condition disappears. For large values of ϵ the constraint is fulfilled more and more exactly but the condition number of the linear equation system (117) will be very large. Papers regarding formulation (117) have been published for the linear case by Malkus and Hughes (1978) and for the large strain case of rubber elasticity by e.g. Häggblad and Sundberg (1983) and Sussman and Bathe (1987).

- **Hu-Washizu functional.** For the fulfillment of the incompressibility constraint in case of finite elastic deformations the functional described in section 1.5 can be employed. It formulates the constraint condition via a constitutive equations for the pressure

$$H(\varphi\,, p\,, \theta\,) = W(\widehat{\boldsymbol{C}}) + K\,[\,G(\theta)]^2 + p\,(J - \theta) \tag{118}$$

Within the finite element discretization ansatz functions are selected for the deformation φ, the pressure p and the volumetric strain θ. $G(\theta)$

[7] This system of incremental equations is equivalent to a penalty formulation for the incompressibility constraint.

defines the constitutive equation for the pressure term, here K is the modulus of compression. The formulation of $W(\widehat{\boldsymbol{C}})$ is provided by (36). The associated discretization within the finite element method was firstly presented in Simo et al. (1985).

Finite elements which are derived from mixed methods have to fulfill additional mathematical conditions which guarantee the stability of the element formulation. This condition is known as BB-condition, named after its inventors Babuska and Brezzi. Its fulfillment is related to the condition that matrix $\boldsymbol{B}_{T\,pu}$ in (116) is not rank deficient.

For general nonlinear applications there exists no formulation of the BB-condition. One can apply the condition for the tangent spaces analogously which belong to a given state of deformation and pressure, as e.g. provided in (114). The BB-condition has the disadvantage that it cannot be formulated for a single element. One always has to consider a patch of elements, see e.g. B Brezzi and Fortin (1991) or Braess (1992). A numerical method to show fulfillment of the BB-condition was derived in Chapelle and Bathe (1993).

Mixed Q1-P0 element for finite deformations In this section a finite element is derived for finite elastic deformations which is based on the Hu-Washizu variational formulation[8]. The element, developed in Simo et al. (1985), is implemented in many existing finite element codes and uses linear shape functions for the deformation field related to the deviatoric kinematical variables. Additionally constant ansatz functions are applied to discretize the pressure and volumetric strain[9].

The continuum mechanical basis for the mixed Q1-P0 element was already discussed in section 1.5. Equation (75) describes the weak form with respect to the spatial configuration. Inserting the finite element approximation into the weak form yields

$$\nabla_x^S \boldsymbol{\eta}_e = \sum_{I=1}^{n} \boldsymbol{B}_{0\,I}\, \boldsymbol{\eta}_I \,. \qquad (119)$$

[8]The same approach can be followed for the development of finite elements based on different nonlinear principles and weak forms.

[9]Of course also higher order approximation can be selected such like Q2-P1 which uses quadratic shape functions for the deformation field and linear interpolation for the pressure and volumetric strain.

Here the matrix $\boldsymbol{B}_{0\,I}$ denotes the standard B-matrix of the linear theory

$$\boldsymbol{B}_{0\,I} = \begin{bmatrix} N_{I,x} & 0 \\ 0 & N_{I,y} \\ N_{I,y} & N_{I,x} \end{bmatrix} .$$

Only the dervatives have to be computed with respect to the current coordinates, see e.g. Wriggers (2008). The virtual volume strain $\operatorname{div} \boldsymbol{\eta}$ occurs in (75) besides the virtual strain. Its discretization leads to

$$\operatorname{div} \boldsymbol{\eta}_e = \sum_{I=1}^{n} \boldsymbol{B}_{V\,I}\, \boldsymbol{\eta}_I \,, \tag{120}$$

with

$$\boldsymbol{B}_{V\,I} =< N_{I\,,1}\,, N_{I\,,2}\,, N_{I\,,2} > . \tag{121}$$

Again the derivatives have to be computed with respect to the current coordinates. Furthermore constant ansatz functions are introduced for the pressure $J\,p = \tau_{vol}$, see (42), and the volume strain θ_e in Ω_e

$$\tau_{vol\,e} = J\,p_e = J\,\bar{p} \qquad \theta_e = \bar{\theta} \tag{122}$$

With these interpolations the weak form (75) can be written as

$$\begin{aligned} D\Pi(\varphi, p, \theta) \cdot \boldsymbol{\eta} &= \bigcup_{e=1}^{n_e} \sum_{I=1}^{n} \boldsymbol{\eta}_I^T \int_{\Omega_e} \{ (\boldsymbol{B}_{0\,I}^T \boldsymbol{\tau}_{iso\,e} + J\, \boldsymbol{B}_{V\,I}^T \bar{p} \} \, d\Omega - \delta P_{EXT} = 0 \\ D\Pi(\varphi, p, \theta)\, \delta p &= \int_{\Omega_e} \delta\bar{p}\, (J_e - \bar{\theta})\, d\Omega = 0 \\ D\Pi(\varphi, p, \theta)\, \delta\theta &= \int_{\Omega_e} \delta\bar{\theta}\, (\frac{\partial W}{\partial \theta} - \bar{p})\, d\Omega = 0 \,. \end{aligned} \tag{123}$$

The integration is performed with respect to the initial configuration. The first equation denotes the weak form of equilibrium where $\boldsymbol{\tau}$ are the Kirchhoff stresses. The second equation is associated with the constraint equation $J_e = \bar{\theta}$ and the the third equation yields the constitutive equation for the pressure $\bar{p}$, see also $(43)_1$. The two last equation in (123) are fulfilled locally on element level since a discontinuous ansatz was used for pressure and volumetric strain. This leads to

$$
\begin{aligned}
\bar{\theta} &= \frac{1}{\Omega_e} \int_{\Omega_e} J_e \, d\Omega = \frac{\varphi(\Omega_e)}{\Omega_e} \\
\bar{p} &= \frac{1}{\Omega_e} \int_{\Omega_e} \frac{\partial W}{\partial \theta} \, d\Omega = \frac{\partial W}{\partial \theta}(\bar{\theta})
\end{aligned} \tag{124}
$$

The discretization of the weak form (75) is now completed and given by equations $(123)_1$ and (124). Note that the volumetric variable θ follows simply from the ratio of the element volume in the current configuration $\varphi(\Omega_e)$ to the element volume in the initial configuration Ω_e.

Linearization of the Q1-P0 element The linearization of (123) yields a matrix form of the Q1-P0 element in which all variables $(\boldsymbol{\varphi}, p, \theta)$ are present. From the first equation of (123) the linearization follows

$$
D^2\Pi \cdot \Delta u = \bigcup_{e=1}^{n_e} \sum_{I=1}^{n} \boldsymbol{\eta}_I^T [\sum_{K=1}^{n} \bar{\boldsymbol{K}}^u_{T_{IK}} \Delta \boldsymbol{u}_K + \boldsymbol{K}^p_{T_I} \Delta \bar{p}] \tag{125}
$$

where the matrices

$$
\begin{aligned}
\bar{\boldsymbol{K}}^u_{T_{IK}} &= \int_{\Omega_e} \Big[(\nabla_{\bar{x}} N_I)^T \, (\bar{p} \, J \, \mathbf{1} + \bar{\boldsymbol{\tau}}_{iso\,e}) \, \nabla_{\bar{x}} N_K \\
&\quad + \bar{\boldsymbol{B}}^T_{0\,I} \, [(\mathbf{1} \otimes \mathbf{1} - 2\mathbb{I}) \, \bar{p} \, J + \mathbb{D}_{iso}] \, \bar{\boldsymbol{B}}_{0\,K} \Big] \, d\Omega \,, \\
\bar{\boldsymbol{K}}^p_{T_I} &= \int_{\Omega_e} \boldsymbol{B}^T_{V\,I} \, J \, d\Omega \,.
\end{aligned} \tag{126}
$$

have been introduced. From the second equation of (123) we derive its linearization with the Jacobi determinant, and the associated discretization, see (120),

$$
\Delta \theta = \frac{1}{\Omega_e} \sum_{K=1}^{n} \int_{\Omega_e} \boldsymbol{B}_{V\,K} \, J \, d\Omega \, \Delta \boldsymbol{u}_K \,. \tag{127}
$$

The third equation of (123) yields the linearization

$$
\Delta \bar{p} = \frac{\partial^2 W}{\partial \theta^2} \Delta \theta \,. \tag{128}
$$

Inserting now (127) in (128) and using this result in (125) and (126) leads to the elimination of the variables for pressure $\Delta \bar{p}$ and volumetric strain $\Delta \theta$ on element level. Thus a pure displacement formulation is obtained which has as tangent stiffness for the element nodes I and K the following expression

$$\begin{aligned}\bar{\boldsymbol{K}}_{T\,IK}^{Q1P0} &= \int\limits_{\Omega_e} \Big[(\nabla_{\bar{x}} N_I)^T \left(\bar{p}\, J\, \mathbf{1} + \bar{\boldsymbol{\tau}}_{iso\,e} \right) \nabla_{\bar{x}} N_K \\ &\quad + \bar{\boldsymbol{B}}_{0\,I}^T \left[(\mathbf{1} \otimes \mathbf{1} - 2\mathbb{I})\, \bar{p}\, J + \mathbb{D}_{iso} \right] \bar{\boldsymbol{B}}_{0\,K} \Big]\, d\Omega \\ &\quad + \frac{1}{\Omega_e} \int\limits_{\Omega_e} \boldsymbol{B}_{V\,I}^T\, J\, d\Omega \left(\frac{\partial^2 W}{\partial \theta^2} \Omega_e \right) \frac{1}{\Omega_e} \int\limits_{\Omega_e} \boldsymbol{B}_{V\,K}\, J\, d\Omega\,. \end{aligned} \tag{129}$$

This element does not fulfill the BB-condition in the geometrical linear theory. Thus it can lead to non stable solutions for the pressure when special loading and boundary conditions are given. Often post processing of the pressures using L^2 smoothing can help. In practical application it has been shown that this element is quite robust for many problems in solid mechanics which quasi-incompressibly material behaviour. Hence it is contained in many commercial finite element codes.

2.3 Mixed finite elements for finite elasticity

Here a mixed finite element formulation is developed which is based on the Neo-Hooke material equation, see Table 4. Due to the fact that it is possible to invert the constitutive equation, as shown section 1.5, the weak formulation (89) related to a Hellinger-Reissner functional can be used as starting point for the finite element development, here written in matrix notation

$$\begin{aligned} D\Pi \cdot \boldsymbol{\eta} &= \int\limits_B \frac{1}{2} \boldsymbol{C}(\boldsymbol{\eta})^T \boldsymbol{S}\, dV - \int\limits_B \boldsymbol{\eta}^T \hat{\boldsymbol{b}}\, dV - \int\limits_{\partial B_\sigma} \boldsymbol{\eta}^T \hat{\boldsymbol{t}}\, dA = 0\,, \\ D\Pi \cdot \boldsymbol{Q} &= \int\limits_B \frac{1}{2} \boldsymbol{Q}^T [\, \boldsymbol{C}(\boldsymbol{u}) - \boldsymbol{f}\,(\boldsymbol{S})\,]\, dA = 0\,. \end{aligned} \tag{130}$$

Interpolation has to be selected for the displacement field $\boldsymbol{u}$ and the stresses $\boldsymbol{S}$. Here we derive a four node quadrilateral based on the isoparametric concept.

For the displacement field and its variation the standard shape functions are used

$$\boldsymbol{u} = \sum_{I=1}^{4} N_I(\xi\,,\eta)\, \boldsymbol{u}_I\,, \qquad \boldsymbol{\eta} = \sum_{I=1}^{4} N_I(\xi\,,\eta)\, \boldsymbol{\eta}_I \tag{131}$$

As usual, the global coordinates are expressed by the same approximation

$$\boldsymbol{X} = \sum_{I=1}^{4} N_I(\xi\,,\eta)\,\boldsymbol{X}_I \tag{132}$$

where the global coordinates $\boldsymbol{X}_I$ define the initial configuration and ξ, η are convective coordinates with regard to the reference element. The interpolation function is defined within the reference element as

$$N_I = \frac{1}{4}\,(\,1 + \xi\,\xi_I\,)\,(\,1 + \eta\,\eta_I\,)\,. \tag{133}$$

Using these ansatz functions the right Cauchy-Green tensor $\boldsymbol{C}$ can be computed at element level via $\boldsymbol{C}_e = \boldsymbol{F}_e^T \boldsymbol{F}_e$ where $\boldsymbol{F}_e$ is given for a two-dimensional element by

$$\boldsymbol{F}_e = \begin{bmatrix} F_{11} & F_{12} \\ F_{21} & F_{22} \end{bmatrix}_e = \sum_{I=1}^{4} \begin{bmatrix} N_{I,X}\,X_I & N_{I,Y}\,X_I \\ N_{I,X}\,Y_I & N_{I,Y}\,Y_I \end{bmatrix} \tag{134}$$

where the derivative with respect to coordinates X and Y are computed by using the standard isoparametric approach. Furthermore the right Cauchy-Green tensor is given at element level by

$$\boldsymbol{C}_e(\boldsymbol{u}) = \begin{Bmatrix} C_{11} \\ C_{22} \\ C_{12} \end{Bmatrix} = \begin{Bmatrix} (F_{11})^2 + (F_{21})^2 \\ (F_{22})^2 + (F_{12})^2 \\ F_{11}\,F_{12} + F_{21}\,F_{22} \end{Bmatrix}. \tag{135}$$

Furthermore the virtual strains related to the Cauchy-Green tensor, $\boldsymbol{C}_e(\boldsymbol{\eta})$, are given by

$$\boldsymbol{C}_e(\boldsymbol{\eta}) = \begin{Bmatrix} \delta C_{11} \\ \delta C_{22} \\ \delta C_{12} \end{Bmatrix} = \sum_{I=1}^{4} \begin{bmatrix} 2\,F_{11}N_{I,X} & 2\,F_{21}N_{I,X} \\ 2\,F_{12}N_{I,Y} & 2\,F_{22}N_{I,Y} \\ F_{11}N_{I,Y} + F_{12}N_{I,X} & F_{21}N_{I,Y} + F_{22}N_{I,X} \end{bmatrix} \boldsymbol{\eta}_I \tag{136}$$

where the components of the deformation gradient F_{ik} are computed using (134).

For the stresses the interpolation introduced by Pian and Sumihara (1984) is chosen. It leads to the matrix form

$$\begin{Bmatrix} S^{\xi\,\xi} \\ S^{\eta\eta} \\ S^{\xi\eta} \end{Bmatrix} = \begin{bmatrix} 1 & 0 & 0 & \eta & 0 \\ 0 & 1 & 0 & 0 & \xi \\ 0 & 0 & 1 & 0 & 0 \end{bmatrix} \begin{Bmatrix} \bar{s}_1 \\ \bar{s}_2 \\ \bar{s}_3 \\ \bar{s}_4 \\ \bar{s}_5 \end{Bmatrix} \tag{137}$$

with respect to the reference element. Note that the stresses components are usually contravariant which is in accordance with the stress power $S^{ik}\,\dot{C}_{ik}$ and the definition of the strain measures via the deformation gradient.

The stresses in the reference element ($\boldsymbol{S} = S^{\alpha\beta}\,\boldsymbol{G}_\alpha \otimes \boldsymbol{G}_\beta$) have now to be transformed to the global coordinate system which is obtained by

$$\begin{aligned} S^{ik} &= \boldsymbol{E}^i \cdot \boldsymbol{S}\,\boldsymbol{E}^k = \boldsymbol{E}_i \cdot (\,S^{\alpha\beta}\,\boldsymbol{G}_\alpha \otimes \boldsymbol{G}_\beta\,)\,\boldsymbol{E}_k \\ &= S^{\alpha\beta}(\boldsymbol{E}_i \cdot \boldsymbol{G}_\alpha)(\boldsymbol{E}_k \cdot \boldsymbol{G}_\beta) \end{aligned}$$

Note that that for orthogonal basis $\boldsymbol{E}^i = \boldsymbol{E}_i$. Furthermore we can define a matrix form of this transformation which is given by $\boldsymbol{S}(\boldsymbol{X}) = \boldsymbol{T}\,\boldsymbol{S}(\boldsymbol{\xi})\,\boldsymbol{T}^T$. In detail we obtain

$$\begin{bmatrix} S^{xx} & S^{xy} \\ S^{yx} & S^{yy} \end{bmatrix} = \begin{bmatrix} T_{11} & T_{12} \\ T_{21} & T_{22} \end{bmatrix} \begin{bmatrix} S^{\xi\xi} & S^{\xi\eta} \\ S^{\eta\xi} & S^{\eta\eta} \end{bmatrix} \begin{bmatrix} T_{11} & T_{21} \\ T_{12} & T_{22} \end{bmatrix} \tag{138}$$

where $T_{i\alpha} = \boldsymbol{E}_i \cdot \boldsymbol{G}_\alpha$.

Since $\boldsymbol{G}_\alpha = \boldsymbol{X}_{,\alpha}$ we can compute it from using the isoparametric interpolation

$$\boldsymbol{G}_\alpha = \sum_{I=1}^{4} N_I(\xi\,,\eta)_{,\alpha}\,\boldsymbol{X}_I\,. \tag{139}$$

Hence

$$T_{i\alpha} = \sum_{I=1}^{4} N_I(\xi\,,\eta)_{,\alpha}\,X_{i\,I} \tag{140}$$

where $X_{i\,I} = \boldsymbol{E}_i \cdot \boldsymbol{X}_I$. Using the interpolation (133) the derivatives are

$$N_{I\,,\xi} = \frac{\xi_I}{4}\,(\,1 + \eta\,\eta_I\,) \qquad N_{I\,,\eta} = \frac{\eta_I}{4}\,(\,1 + \xi\,\xi_I\,)\,. \tag{141}$$

Within the interpolation of the stresses the transformation matrix will be evaluated at the element center $\xi = \eta = 0$ leading to

$$N^0_{I\,,\xi} = \frac{\xi_I}{4} \qquad N^0_{I\,,\eta} = \frac{\eta_I}{4}\,. \tag{142}$$

and

$$T^0_{i\xi} = \sum_{I=1}^{4} \frac{\xi_I}{4}\,X_{i\,I} \quad \text{and} \quad T^0_{i\eta} = \sum_{I=1}^{4} \frac{\eta_I}{4}\,X_{i\,I}\,. \tag{143}$$

Due to this the transformation matrix is

$$\boldsymbol{T}^0 = \frac{1}{4}\sum_{I=1}^{4} \begin{bmatrix} \xi_I\,X_{1\,I} & \eta_I\,X_{1\,I} \\ \xi_I\,X_{2\,I} & \eta_I\,X_{2\,I} \end{bmatrix} \tag{144}$$

By performing the multiplication in (138) and rearranging to Voigt notation the stress transformation can be written as

$$\begin{Bmatrix} S^{xx} \\ S^{yy} \\ S^{xy} \end{Bmatrix} = \begin{bmatrix} T_{11}^2 & T_{12}^2 & 2\,T_{11}\,T_{12} \\ T_{21}^2 & T_{22}^2 & 2\,T_{21}\,T_{22} \\ T_{11}\,T_{21} & T_{22}\,T_{12} & T_{12}\,T_{21} + T_{11}\,T_{22} \end{bmatrix} \begin{Bmatrix} S^{\xi\xi} \\ S^{\eta\eta} \\ S^{\xi\eta} \end{Bmatrix} \tag{145}$$

Since the stress interpolation (137) is constant for the shear stresses and the transformation matrix is constant element wise, one can find a different representation using the element wise constant matrix $\{\, s_1 \,, s_2 \,, s_3 \,, s_4 \,, s_5 \,\}$. This ansatz can be written in global coordinates using the transformation (145) at the mid point of the element and leads to the simpler form

$$\begin{Bmatrix} S^{xx} \\ S^{yy} \\ S^{xy} \end{Bmatrix} = \begin{Bmatrix} s_1 \\ s_2 \\ s_3 \end{Bmatrix} + \begin{bmatrix} \eta\,(T_{11}^0)^2 & \xi\,(T_{12}^0)^2 \\ \eta\,(T_{21}^0)^2 & \xi\,(T_{22}^0)^2 \\ \eta\,T_{11}^0\,T_{21}^0 & \xi\,T_{22}^0\,T_{12}^0 \end{bmatrix} \begin{Bmatrix} s_4 \\ s_5 \end{Bmatrix} . \tag{146}$$

The interpolations for displacements and stresses can now be used within the mixed weak form (130) and its linearization to derive the matrix form of the finite element formulation.

2.4 Example

By means of one example we like to show the behaviour of different finite elements. As a classical example for benchmarking two-dimensional finite elements in finite strain situations the Cook's membrane problem was selected. The system and data are is depicted in Figure 2. The material is assumed to be hyperelastic and incompressible. Hence only elements which perform well for incompressibility and bending can produce good results.

Compared are the standard Q1 element against the Q1/P0 element derived in section 2.2 and the element developed in section 2.3 which is called PS8. Furthermore results obtained with the CB/E4 element, derived in Korelc and Wriggers (1996) are shown in Fig. 3. It depicts the convergence behaviour of the different element formulations. It is found that all elements which are developed based on mixed methods yield very good results.

However when robustness of the element formulation is concerned, then the PS8 element seems to be superior. This can be observed from Table 5 in which the tip displacement for different mesh sizes is reported. Observe that the Q1 displacement element in the first line of this table produces a tip displacement with a 128×128 mesh which has the same accuracy as the results obtained with a 2×2 mesh when using the mixed elements hence the Q1 element locks severely in this test. Note further that the PS8 element can reach the final displacement in only one load step $\Delta F = 100$ while the

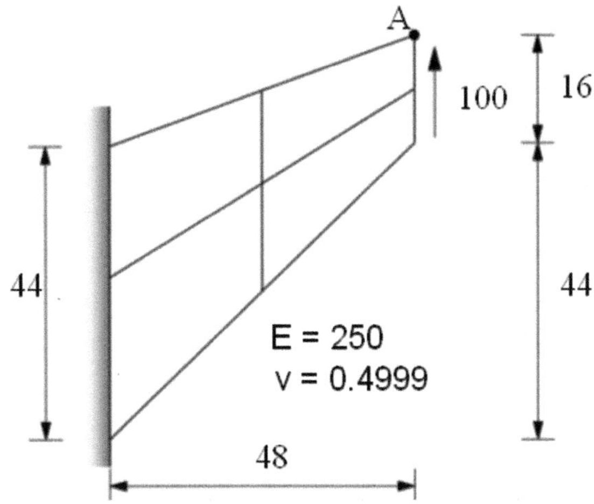

Figure 2. Cook's membran problem, incompressible

enhanced element Q1/E4 and CG4 need 20 load steps to obtain the final tip displacement. This result underlines the robustness of the PS8 element formulation which was one demand for user friendly numerical simulations of nonlinear problems. However the formulation of the PS8 element is complex and this element cannot be applied to general constitutive equations.

Bibliography

D. N. Arnold, F. Brezzi, and J. Douglas. Peers: A new mixed finite element for plane elasticity. *Japan Journal of Applied Mathematics*, 1:347–367, 1984.

K. J. Bathe. *Finite Element Procedures.* Prentice-Hall, Englewood Cliffs, New Jersey, 1996.

K. J. Bathe, E. Ramm, and E. L. Wilson. Finite element formulation for large deformation analysis. *International Journal for Numerical Methods in Engineering*, 9:353–386, 1975.

T. Belytschko, J. S.-J. Ong, W. K. Liu, and J. M. Kennedy. Hourglass control in linear and nonlinear problems. *Computer Methods in Applied Mechanics and Engineering*, 43:251–276, 1984.

D. Braess. *Finite Elemente.* Springer-Verlag, Berlin, Heidelberg, New York, 1992.

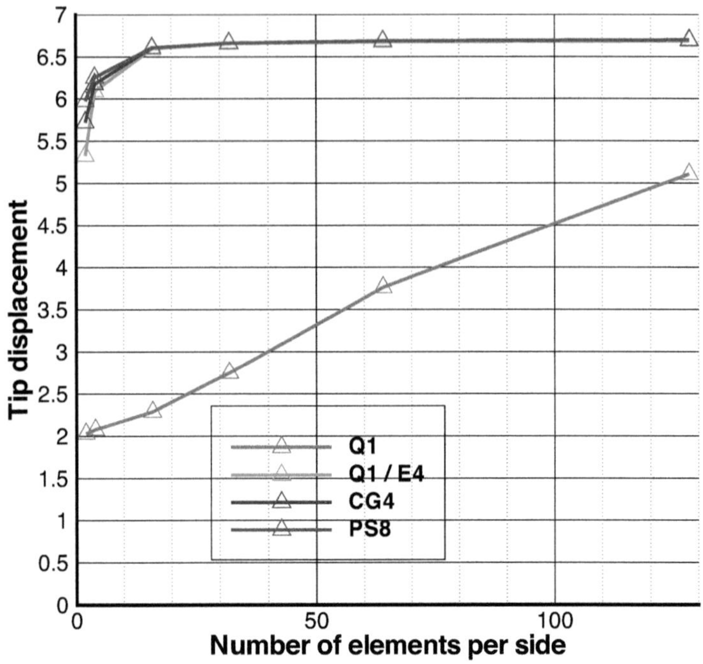

Figure 3. Convergence of the different element formulations

F. Brezzi and M. Fortin. *Mixed and hybrid finite element Methods.* Springer, Berlin, Heidelberg, New York, 1991.

P. Chadwick. *Continuum Mechanics, Concise Theory and Problems.* Dover Publications, Mineola, 1999.

D. Chapelle and K. J. Bathe. The inf-sup test. *Computers and Structures*, 47:537–545, 1993.

P. G. Ciarlet. *Mathematical Elasticity I: Three-dimensional Elasticity.* North-Holland, Amsterdam, 1988.

G. Duffet and B. D. Reddy. The Analysis of Incompressible Hyperelastic Bodies by the Finite Element Method. *Computer Methods in Applied Mechanics and Engineering*, 41:105–120, 1983.

	ΔF	2 x 2	4 x 4	16 x 16	32 x 32	128 x 128
disp.	50.	2.034	2.082	2.292	2.752	5.111
Q1E4	5.	5.332	6.096	6.600	6.661	
PS8	100.	5.978	6.259	6.610	6.665	6.702
QB/E4	5.	5.999	6.260	6.611	6.662	

Table 5. Convergence of the different element formulations

P.J. Flory. Thermodynamic relations for high elastic materials. *Trans. Faraday. Soc.*, 57:829–838, 1961.

B. M. Fraeijs de Veubeke. Stress function approach. In *World Congress on the Finite Element Method in Structural Mechanics*, pages 1–51. Bournmouth, 1975.

S. Glaser and F. Armero. On the Formulation of Enhanced Strain Finite Elements in Finite Deformations. *Engineering Computations*, 14:759–791, 1997.

F. Gruttmann, W. Wagner, and P. Wriggers. A nonlinear quadrilateral shell element with drilling degrees of freedom. *Ingenieur Archiv*, 62:474–486, 1992.

B. Häggblad and J. A. Sundberg. Large strain solutions of rubber components. *Computers and Structures*, 17:835–843, 1983.

G. A. Holzapfel. *Nonlinear Solid Mechanics*. Wiley, Chichester, 2000.

T. J. R. Hughes. Generalization of selective integration procedures to anisotropic and nonlinear media. *International Journal for Numerical Methods in Engineering*, 15:1413–1418, 1980.

T. J. R. Hughes and F. Brezzi. On drilling degrees of freedom. *Computer Methods in Applied Mechanics and Engineering*, 72:105–121, 1989.

T. R. J. Hughes. *The Finite Element Method.* Prentice Hall, Englewood Cliffs, New Jersey, 1987.

A. Ibrahimbegovic, R. L. Taylor, and E. L. Wilson. A robust quadrilateral membrane element with drilling degrees of freedom. *International Journal for Numerical Methods in Engineering*, 30:445–457, 1990.

M. Iura and S. N. Atluri. Formulation of a membrane finite element with drilling degrees of freedom. *Computational Mechanics*, 39:417–428, 1992.

J. Korelc and P. Wriggers. Consistent gradient formulation for a stable enhanced strain method for large deformations. *Engineering Computations*, 13:103–123, 1996.

D. S. Malkus and T. J. R. Hughes. Mixed finite element methods - reduced and selective integration techniques: a unification of concepts. *Computer Methods in Applied Mechanics and Engineering*, 15:63–81, 1978.

L. E. Malvern. *Introduction to the Mechanics of a Continuous Medium.* Prentice-Hall, Englewood Cliffs, New Jersey, 1969.

J. E. Marsden and T. J. R. Hughes. *Mathematical Foundations of Elasticity.* Prentice-Hall, Englewood Cliffs, New Jersey, 1983.

J. T. Oden and J. E. Key. Numerical analysis of finite axisymmetrical deformations of incompressible elastic solids of revolution. *International Journal of Solids & Structures*, 6:497–518, 1970.

R. W. Ogden. *Non-Linear Elastic Deformations.* Ellis Horwood and John Wiley, Chichester, 1984.

T. H. H. Pian. Derivation of element stiffness matrices by assumed stress distributions. *AIAA-J. 2*, 7:1333–1336, 1964.

T. H. H. Pian and K. Sumihara. Rational approach for assumed stress finite elements. *International Journal for Numerical Methods in Engineering*, 20:1685–1695, 1984.

S. Reese. On a physically stabilized one point finite element formulation for three-dimensional finite elasto-plasticity. *Computer Methods in Applied Mechanics and Engineering*, 194:4685–4715, 2005.

S. Reese and P. Wriggers. A new stabilization concept for finite elements in large deformation problems. *International Journal for Numerical Methods in Engineering*, 48:79–110, 2000.

O. Schenk and K. Gärtner. Solving unsymmetric sparse systems of linear equations with pardiso. *Journal of Future Generation Computer Systems*, 20:475–487, 2004.

J. C. Simo and F. Armero. Geometrically Non-Linear Enhanced Strain Mixed Methods and the Method of Incompatible Modes. *International Journal for Numerical Methods in Engineering*, 33:1413–1449, 1992.

J. C. Simo and T. J. R. Hughes. *Computational Inelasticity.* Springer, New York, Berlin, 1998.

J. C. Simo and M. S. Rifai. A class of assumed strain methods and the method of incompatible modes. *International Journal for Numerical Methods in Engineering*, 29:1595–1638, 1990.

J. C. Simo and R. L. Taylor. Quasi-incompressible finite elasticity in principal stretches. continuum basis and numerical algorithms. *Computer Methods in Applied Mechanics and Engineering*, 85:273–310, 1991.

J. C. Simo, R. L. Taylor, and K. S. Pister. Variational and projection methods for the volume constraint in finite deformation elasto-plasticity. *Computer Methods in Applied Mechanics and Engineering*, 51:177–208, 1985.

J. C. Simo, F. Armero, and R. L. Taylor. Improved Versions of Assumed Enhanced Strain Tri-Linear Elements for 3D Finite Deformation Problems. *Computer Methods in Applied Mechanics and Engineering*, 110: 359–386, 1993.

R. Stenberg. A family of mixed finite elements for elasticity problems. *Numerische Mathematik*, 48:513–538, 1988.

T. Sussman and K.-J. Bathe. A Finite Element Formulation for Nonlinear Incompressible Elastic and Inelastic Analysis. *Computers and Structures*, 26:357–409, 1987.

R. L. Taylor, P. J. Beresford, and E. L. Wilson. A Non-Conforming Element for Stress Analysis. *International Journal for Numerical Methods in Engineering*, 10:1211–1219, 1976.

C. Truesdell and W. Noll. The nonlinear field theories of mechanics. In S. Flügge, editor, *Handbuch der Physik III/3*. Springer, Berlin, Heidelberg, Wien, 1965.

C. Truesdell and R. Toupin. The classical field theories. In *Handbuch der Physik III/1*. Springer, Berlin, Heidelberg, Wien, 1960.

K. Washizu. *Variational Methods in Elasticity and Plasticity*. Pergamon Press, Oxford, second edition, 1975.

E. L. Wilson, R. L. Taylor, W. P. Doherty, and J. Ghaboussi. Incompatible Displacements Models. In *Numerical and Computer Models in Structural Mechanics*, Fenves S. J., Perrone N., Robinson A. R. and Schnobrich W. C. (Eds.), New York, 1973. Academic Press. 43–57.

P. Wriggers. *Nichtlineare Finite Elemente*. Springer, Berlin, 2001.

P. Wriggers. *Nonlinear Finite Elements*. Springer, Berlin, Heidelberg, New York, 2008.

P. Wriggers and S. Reese. A note on enhanced strain methods for large deformations. *Computer Methods in Applied Mechanics and Engineering*, 135:201–209, 1996.

O. C. Zienkiewicz and R. L. Taylor. *The Finite Element Method, 4th Ed.*, volume 1. McGraw Hill, London, 1989.

O. C. Zienkiewicz, R. L. Taylor, and J. M. Too. Reduced Integration Technique in General Analysis of Plates and Shells. *International Journal for Numerical Methods in Engineering*, 3:275–290, 1971.

Stability of Some Finite Element Methods for Finite Elasticity Problems

F. Auricchio [*♯§], L. Beirão da Veiga [‡♯], C. Lovadina [†♯], and A. Reali [*§]

[*] Dipartimento di Meccanica Strutturale, Università di Pavia, Italy
[†] Dipartimento di Matematica, Università di Pavia, Italy
[‡] Dipartimento di Matematica, Università di Milano, Italy
[♯] IMATI – CNR, Pavia, Italy
[§] ROSE School, Università di Pavia, Italy

Abstract We consider the finite elasticity problem for incompressible materials, proposing a simple bidimensional problem for which we provide indications on the solution stability. Furthermore, we study the stability of the discrete solution, obtained by means of some well-known finite elements, and we present several numerical experiments in order to evaluate and compare the performance of the different discrete schemes under investigation on the considered model problem.

1 Introduction

Nowadays there are several finite element interpolation schemes performing very well, in terms of both accuracy and stability, for the case of small deformation problems, also in the presence of highly constrained situations (i.e. incompressible materials). Examples with excellent performance ranges from standard mixed elements (see, for instance, Bathe, 1996, Brezzi and Fortin, 1991, Hughes, 2000, and the references therein) to enhanced strain elements (see Auricchio et al., 2005a, Braess, 1998, Pantuso and Bathe, 1995, Reddy and Simo, 1995, Simo and Rifai, 1990).

However, it is also well established that the extension of such schemes to the case of finite strain problems is by no way trivial; in particular, even elements which seem to be ideal from a theoretical and a numerical perspective may fail in the large strain range, for example due to the rising of unphysical instabilities for high compression/tension states (see Armero, 2000, Auricchio et al., 2005b, Lovadina and Auricchio, 2003, Pantuso and Bathe, 1997, Wriggers and Reese, 1996, for more details). It is worth recalling that many interesting strategies have been developed in order to stabilize the methods at hand (cf., for instance, Klaas et al., 1999, Maniatty et al., 2002, Nagtegaal and Fox, 1996, Reese et al., 1999, Reese and Wriggers, 2000, Simo and Armero, 1992). However, a satisfactory analysis of finite element methods for finite strain problems is still missing.

According to the cited problems, this work focuses on a simple finite-strain elastic bidimensional problem for which not only it is possible to compute the solution in closed form but it is also possible to draw some indications on the solution stability.

After recalling the general finite strain elasticity framework (Section 2), we present our 2D continuum problem (Section 3). Essentially, we consider an incompressible square block clamped on three sides, and subjected to a uniaxial force density, for which we vary the intensity by introducing a real parameter. For all parameter values the proposed problem admits a trivial solution, for which we study the (linearized) stability. We thus identify a *stability range* for the parameter choice in our continuum model problem (Section 3.1). We then present some possible finite element discretizations of the problem under investigation. In particular, in some cases, we are able to present theoretical results on the stability ranges of the *discrete* problems, showing the capability (or the incapability) of the considered elements to reproduce the continuum stability features (Section 4). We then perform extensive numerical simulations to investigate the whole stability range of such finite element discretizations (Section 5). We conclude by comparing the performance of the different discrete schemes with reference to the stability range expected for the continuous problem. Therefore, we can identify the elements which promise to be most reliable for the analysis of large deformation problems.

We remark that most of the results presented in this work have already been discussed in Auricchio et al. (2005b).

2 The Finite Strain Incompressible Elasticity Problem

In this work we adopt the so-called *material description* to study the finite strain elasticity problem. Accordingly, we suppose that we are given a *reference configuration* $\Omega \subset \mathcal{R}^d$ for a d-dimensional bounded material body $\mathcal{B}$. Therefore, the deformation of $\mathcal{B}$ can be described by means of the map $\hat{\varphi} : \Omega \to \mathcal{R}^d$ defined by

$$\hat{\varphi}(\mathbf{X}) = \mathbf{X} + \hat{\mathbf{u}}(\mathbf{X}) \ , \tag{1}$$

where $\mathbf{X} = (X_1, .., X_d)$ denote the coordinates of a material point in the reference configuration and $\hat{\mathbf{u}}(\mathbf{X})$ represents the corresponding displacement vector. Following standard notations, we introduce the deformation gradient $\hat{\mathbf{F}} = \mathbf{F}(\hat{\mathbf{u}})$ and the right Cauchy-Green deformation tensor $\hat{\mathbf{C}} = \mathbf{C}(\hat{\mathbf{u}})$ by setting

$$\hat{\mathbf{F}} = \mathbf{I} + \nabla\hat{\mathbf{u}} \quad , \quad \hat{\mathbf{C}} = \hat{\mathbf{F}}^T\hat{\mathbf{F}} \, , \tag{2}$$

where $\mathbf{I}$ is the second-order identity tensor and ∇ is the gradient operator with respect to the coordinates $\mathbf{X}$.

For a homogeneous neo-Hookean material we define (see for example Bonet and Wood, 1997, and Ciarlet, 1978) the potential energy function as

$$\Psi(\hat{\mathbf{u}}) = \frac{1}{2}\mu\left[\mathbf{I} : \hat{\mathbf{C}} - d\right] - \mu \ln \hat{J} + \frac{\lambda}{2}\,\Theta(\hat{J})^2 \, , \tag{3}$$

where λ and μ are positive constants, " : " represents the usual inner product for second-order tensors and $\hat{J} = \det \hat{\mathbf{F}}$. Moreover, Θ is a real function usually chosen as

$$\Theta(J) = \ln J \qquad \text{or} \qquad \Theta(J) = J - 1 \, . \tag{4}$$

Introducing the pressure-like variable (or simply pressure) $\hat{p} = \lambda\Theta(\hat{J})$, the potential energy (3) can be equivalently written as the following function of $\hat{\mathbf{u}}$ and $\hat{p}$ (still denoted with Ψ, with a little abuse of notations)

$$\Psi(\hat{\mathbf{u}}, \hat{p}) = \frac{1}{2}\mu\left[\mathbf{I} : \hat{\mathbf{C}} - d\right] - \mu \ln \hat{J} + \hat{p}\Theta(\hat{J}) - \frac{1}{2\lambda}\hat{p}^2 \, . \tag{5}$$

When the body $\mathcal{B}$ is subjected to a given load $\mathbf{b} = \mathbf{b}(\mathbf{X})$ per unit volume in the reference configuration, the total elastic energy functional reads as follows

$$\Pi(\hat{\mathbf{u}}, \hat{p}) = \int_\Omega \Psi(\hat{\mathbf{u}}, \hat{p}) - \int_\Omega \mathbf{b} \cdot \hat{\mathbf{u}} \, . \tag{6}$$

Therefore, following the Hellinger-Reissner variational principle, equilibrium is derived by searching for critical points of (6) in suitable admissible displacement and pressure spaces $\hat{U}$ and $\hat{P}$. The corresponding Euler-Lagrange equations emanating from (6) lead to solve

$$\begin{cases} \text{Find } (\hat{\mathbf{u}}, \hat{p}) \in \hat{U} \times \hat{P} \text{ such that} \\[2ex] \mu \displaystyle\int_\Omega [\hat{\mathbf{F}} - \hat{\mathbf{F}}^{-T}] : \nabla\mathbf{v} + \int_\Omega \hat{p}\,\Theta'(\hat{J})\hat{J}\,\hat{\mathbf{F}}^{-T} : \nabla\mathbf{v} = \int_\Omega \mathbf{b} \cdot \mathbf{v} \qquad \forall \mathbf{v} \in U \\[2ex] \displaystyle\int_\Omega \left(\Theta(\hat{J}) - \frac{\hat{p}}{\lambda}\right) q = 0 \qquad \forall q \in P \, , \end{cases} \tag{7}$$

where U and P are the admissible variation spaces for the displacements and the pressures, respectively. We note that in (7) we used that the linearization of the deformation gradient jacobian is

$$DJ(\hat{\mathbf{u}})[\mathbf{v}] = J(\hat{\mathbf{u}})\mathbf{F}(\hat{\mathbf{u}})^{-T} : \nabla\mathbf{v} = \hat{J}\hat{\mathbf{F}}^{-T} : \nabla\mathbf{v} \qquad \forall \mathbf{v} \in U \; . \tag{8}$$

Without loss of generality, from now on we select $\Theta(J) = \ln J$ (see (4)). Moreover, we focus on the case of an *incompressible* material, which corresponds to take the limit $\lambda \to +\infty$ in (7). Therefore, our problem becomes

$$\begin{cases} \text{Find } (\hat{\mathbf{u}}, \hat{p}) \in \hat{U} \times \hat{P} \text{ such that} \\[2ex] \mu \displaystyle\int_\Omega \hat{\mathbf{F}} : \nabla\mathbf{v} + \int_\Omega (\hat{p} - \mu)\hat{\mathbf{F}}^{-T} : \nabla\mathbf{v} - \int_\Omega \mathbf{b}\cdot\mathbf{v} = \mathbf{0} \qquad \forall \mathbf{v} \in U \\[2ex] \displaystyle\int_\Omega q \, \ln \hat{J} = 0 \qquad \forall q \in P \; , \end{cases} \tag{9}$$

or, in residual form,

$$\begin{cases} \text{Find } (\hat{\mathbf{u}}, \hat{p}) \in \hat{U} \times \hat{P} \text{ such that} \\[2ex] \mathcal{R}_u((\hat{\mathbf{u}}, \hat{p}), \mathbf{v}) = \mathbf{0} \qquad \forall \mathbf{v} \in U \\[2ex] \mathcal{R}_p((\hat{\mathbf{u}}, \hat{p}), q) = 0 \qquad \forall q \in P \; , \end{cases} \tag{10}$$

where

$$\begin{cases} \mathcal{R}_u((\hat{\mathbf{u}}, \hat{p}), \mathbf{v}) := \mu \displaystyle\int_\Omega \hat{\mathbf{F}} : \nabla\mathbf{v} + \int_\Omega (\hat{p} - \mu)\hat{\mathbf{F}}^{-T} : \nabla\mathbf{v} - \int_\Omega \mathbf{b}\cdot\mathbf{v} \\[2ex] \mathcal{R}_p((\hat{\mathbf{u}}, \hat{p}), q) := \displaystyle\int_\Omega q \, \ln \hat{J} \; . \end{cases} \tag{11}$$

We now derive the linearization of problem (9) around a generic point $(\hat{\mathbf{u}}, \hat{p})$. Observing that

$$D\hat{\mathbf{F}}^{-T}(\hat{\mathbf{u}})[\mathbf{u}] = -\hat{\mathbf{F}}^{-T}(\nabla\mathbf{u})^T\hat{\mathbf{F}}^{-T} \qquad \forall \mathbf{u} \in U \; . \tag{12}$$

we easily get the problem for the *infinitesimal* increment $(\mathbf{u}, p)$

$$
\begin{cases}
\text{Find } (\mathbf{u}, p) \in U \times P \text{ such that} \\[2ex]
\mu \int_\Omega \nabla \mathbf{u} : \nabla \mathbf{v} + \int_\Omega (\mu - \hat{p})(\hat{\mathbf{F}}^{-1}\nabla \mathbf{u})^T : \hat{\mathbf{F}}^{-1}\nabla \mathbf{v} \\[2ex]
\qquad + \int_\Omega p\hat{\mathbf{F}}^{-T} : \nabla \mathbf{v} = -\mathcal{R}_u((\hat{\mathbf{u}}, \hat{p}), \mathbf{v}) \qquad \forall \mathbf{v} \in U \\[2ex]
\int_\Omega q\hat{\mathbf{F}}^{-T} : \nabla \mathbf{u} = -\mathcal{R}_p((\hat{\mathbf{u}}, \hat{p}), q) \qquad \forall q \in P \, .
\end{cases}
\tag{13}
$$

Remark 2.1. Since problem (13) is the linearization of problem (9) (or equivalently (10)), it can be interpreted as the generic step of a Newton-like iteration procedure for the solution of the non-linear problem (9).

Remark 2.2. Taking $(\hat{\mathbf{u}}, \hat{p}) = (0, 0)$ in (13), we immediately recover the classical linear incompressible elasticity problem for small deformations, i.e.

$$
\begin{cases}
\text{Find } (\mathbf{u}, p) \in U \times P \text{ such that} \\[2ex]
2\mu \int_\Omega \boldsymbol{\varepsilon}(\mathbf{u}) : \boldsymbol{\varepsilon}(\mathbf{v}) + \int_\Omega p \, \mathrm{div}\mathbf{v} = \int_\Omega \mathbf{b} \cdot \mathbf{v} \qquad \forall \mathbf{v} \in U \\[2ex]
\int_\Omega q \, \mathrm{div}\mathbf{u} = 0 \qquad \forall q \in P \, ,
\end{cases}
\tag{14}
$$

where $\boldsymbol{\varepsilon}(\cdot)$ denotes the symmetric gradient operator.

3 A Model Problem for Finite Strain Incompressible elasticity

In this Section we present a simple bidimensional problem which nonetheless shows some of the difficulties arising in general nonlinear elastic problems for incompressible materials.

Using the usual Cartesian coordinates (X, Y), we consider a square material body whose reference configuration is $\Omega = (-1, 1) \times (-1, 1)$; we denote with $\Gamma = [-1, 1] \times \{1\}$ the upper part of its boundary, while the remaining part of $\partial\Omega$ is denoted with Γ_D. The body Ω is clamped along Γ_D and subjected to the volume force $\mathbf{b} = \gamma \mathbf{f}$, where $\mathbf{f} = (0,1)^T$ and γ is a real parameter (cf. Fig. 1).

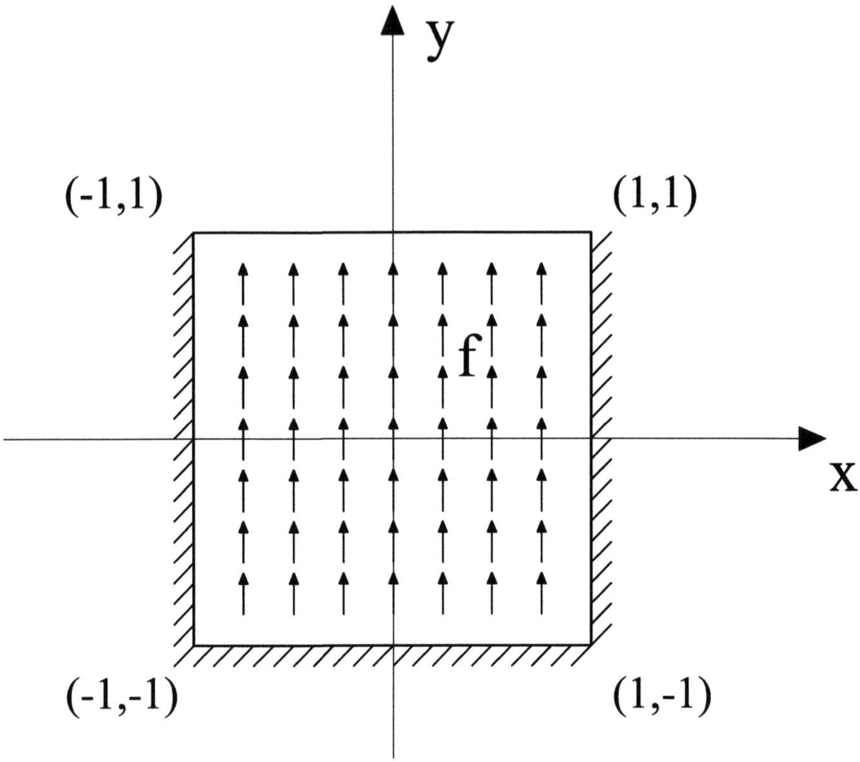

Figure 1. Problem 1: square incompressible block clamped on three edges and subjected to a vertical uniform body load.

Therefore, the equilibrium problem leads to solve the following variational system (see (9))

$$\begin{cases} \text{Find } (\hat{\mathbf{u}}, \hat{p}) \in \hat{U} \times \hat{P} \text{ such that} \\[2ex] \mu \displaystyle\int_\Omega \hat{\mathbf{F}} : \nabla \mathbf{v} + \int_\Omega (\hat{p} - \mu) \hat{\mathbf{F}}^{-T} : \nabla \mathbf{v} = \gamma \int_\Omega \mathbf{f} \cdot \mathbf{v} \qquad \forall \mathbf{v} \in U \\[2ex] \displaystyle\int_\Omega q \ln \hat{J} = 0 \qquad \forall q \in P\,. \end{cases} \tag{15}$$

It is not our intention to rigorously specify the regularity needed for the space involved in the variational formulation (15), and we refer to Le Tallec (1982, 1994) for details on such a point. Instead, we wish to notice that system (15) constitutes a set of *nonlinear* equations for which a trivial solution can be easily found *for every* $\gamma \in \boldsymbol{R}$, i.e. $(\hat{\mathbf{u}}, \hat{p}) = (\mathbf{0}, \gamma r)$, where $r = r(X, Y) = 1 - Y$.

Remark 3.1. We are not claiming that, for each $\gamma \in \mathbf{R}$, $(\hat{\mathbf{u}}, \hat{p}) = (\mathbf{0}, \gamma r)$ is the only solution of the system.

Whenever an incremental loading procedure is considered, the passage from γ to $\gamma + \Delta\gamma$ in (15) is typically solved by a Newton's technique. Supposing that at γ convergence has been reached, the first iteration step of Newton's method with initial guess $(\hat{\mathbf{u}}, \hat{p}) = (\mathbf{0}, \gamma r)$ consists (recalling the linearized problem (13)) in solving

$$\begin{cases} \text{Find } (\mathbf{u}, p) \in U \times P \text{ such that} \\[2ex] 2\mu \displaystyle\int_\Omega \boldsymbol{\varepsilon}(\mathbf{u}) : \boldsymbol{\varepsilon}(\mathbf{v}) - \gamma \int_\Omega r (\nabla \mathbf{u})^T : \nabla \mathbf{v} + \int_\Omega p \mathrm{div}\mathbf{v} = \Delta\gamma \int_\Omega \mathbf{f} \cdot \mathbf{v} \\[2ex] \displaystyle\int_\Omega q \mathrm{div}\mathbf{u} = 0 \ , \end{cases} \tag{16}$$

for every $(\mathbf{v}, q) \in U \times P$. Letting

$$U = \left\{ \mathbf{v} \in H^1(\Omega)^2 \ : \ \mathbf{v}_{|\Gamma_D} = 0 \right\} \ ; \qquad P = L^2(\Omega) \ , \tag{17}$$

we consider system (16) as a *model problem* for our subsequent considerations. Denoting with $\mathbf{A}^S$ the symmetric part of a generic second-order tensor $\mathbf{A}$, introducing the bilinear forms

$$a_\gamma(\mathbf{F}, \mathbf{G}) =: 2\mu \int_\Omega \mathbf{F}^S : \mathbf{G}^S - \gamma \int_\Omega r \, \mathbf{F}^T : \mathbf{G} \tag{18}$$

and

$$b(\mathbf{v}, q) =: \int_\Omega q \, \mathrm{div}\mathbf{v} \ , \tag{19}$$

problem (16) can be written as

$$\begin{cases} \text{Find } (\mathbf{u}, p) \in U \times P \text{ such that} \\ \\ a_\gamma(\nabla \mathbf{u}, \nabla \mathbf{v}) + b(\mathbf{v}, p) = \Delta\gamma \int_\Omega \mathbf{f} \cdot \mathbf{v} \qquad \forall \mathbf{v} \in U \\ \\ b(\mathbf{u}, q) = 0 \qquad \forall q \in P \ . \end{cases} \tag{20}$$

Therefore, we are clearly facing a typical (parameter-depending) saddle-point problem. As it is well-established (cf. Brezzi and Fortin, 1991), the crucial properties for the well-posedness are, together with continuity:

- *the inf-sup condition*, i.e. the existence of a positive constant β such that
$$\inf_{q \in P} \sup_{\mathbf{v} \in U} \frac{b(\mathbf{v}, q)}{||\mathbf{v}||_U ||q||_P} \geq \beta \ ; \tag{21}$$
- *the invertibility on the kernel condition*, i.e. the existence of a positive constant $\alpha(\gamma, \mu)$ such that
$$\inf_{\mathbf{v} \in Ker\, B} \sup_{\mathbf{u} \in Ker\, B} \frac{a_\gamma(\nabla \mathbf{u}, \nabla \mathbf{v})}{||\mathbf{u}||_U ||\mathbf{v}||_U} \geq \alpha(\gamma, \mu) \ , \tag{22}$$
where
$$Ker\, B = \{\mathbf{v} \in U \ : \ b(\mathbf{v}, q) = 0 \qquad \forall q \in P\} \ . \tag{23}$$

As far as the inf-sup condition is concerned, it is a classical result that it holds for the divergence operator. We therefore focus our attention on condition (22). In particular, we will show that the form $a_\gamma(\cdot, \cdot)$ is *coercive* on $Ker\, B$ whenever γ stays in a suitable range of values. We thus expect that within such choices of the parameters, for the continuous problem the trivial solution $(\hat{\mathbf{u}}, \hat{p}) = (\mathbf{0}, \gamma r)$ is unique and stable.

3.1 The Stability Range

We now investigate on the coercivity on $Ker\, B$ of $a_\gamma(\cdot, \cdot)$. More precisely, recalling the well-known Korn's inequality, we search for conditions on γ implying the existence of a constant $c(\gamma, \mu) > 0$ such that

$$2\mu \int_\Omega |\boldsymbol{\varepsilon}(\mathbf{v})|^2 - \gamma \int_\Omega r\, (\nabla \mathbf{v})^T : \nabla \mathbf{v} \geq c(\gamma, \mu) \int_\Omega |\boldsymbol{\varepsilon}(\mathbf{v})|^2 \qquad \forall \mathbf{v} \in Ker\, B \ . \tag{24}$$

Since $\text{div}(\nabla\mathbf{v})^T = \mathbf{0}$ for every divergence-free function $\mathbf{v}$ (i.e. for every $\mathbf{v} \in Ker\, B$), an integration by parts gives

$$2\mu \int_\Omega |\boldsymbol{\varepsilon}(\mathbf{v})|^2 - \gamma \int_\Omega r\,(\nabla\mathbf{v})^T : \nabla\mathbf{v} = 2\mu \int_\Omega |\boldsymbol{\varepsilon}(\mathbf{v})|^2 + \gamma \int_\Omega \nabla r \cdot [\nabla\mathbf{v}]\mathbf{v} \,. \quad (25)$$

Above, the boundary integral arising from integration by parts disappears because of the boundary conditions on $\mathbf{v}$, and because on Γ the function $r = 1 - Y$ vanishes.

Now, a further integration by parts leads to

$$\gamma \int_\Omega \nabla r \cdot [\nabla\mathbf{v}]\mathbf{v} = \gamma \left(- \int_\Omega H(r)\mathbf{v}\cdot\mathbf{v} + \int_\Gamma (\mathbf{v}\cdot\nabla r)(\mathbf{v}\cdot\mathbf{n}) \right) , \quad (26)$$

where $\mathbf{n}$ is the outward normal vector and $H(r)$ is the Hessian matrix of the function r. But $r = 1 - Y$ is linear, hence $H(r) = 0$. On the other hand, on the boundary Γ we have $\nabla r = -\mathbf{n}$, so that we deduce that

$$\gamma \int_\Omega \nabla r \cdot [\nabla\mathbf{v}]\mathbf{v} = -\gamma \int_\Gamma (\mathbf{v}\cdot\mathbf{n})^2 \,. \quad (27)$$

From (24), (25) and (27) we conclude that our form $a_\gamma(\cdot,\cdot)$ will be coercive on $Ker\, B$ if there exists a constant $c(\gamma,\mu) > 0$ such that

$$2\mu \int_\Omega |\boldsymbol{\varepsilon}(\mathbf{v})|^2 - \gamma \int_\Gamma (\mathbf{v}\cdot\mathbf{n})^2 \geq c(\gamma,\mu) \int_\Omega |\boldsymbol{\varepsilon}(\mathbf{v})|^2 \qquad \forall \mathbf{v} \in Ker\, B \,. \quad (28)$$

By (28) we first infer that for $\gamma \leq 0$ we can simply take $c(\gamma,\mu) = 2\mu$. Furthermore, setting

$$\alpha_M = \sup_{\mathbf{v}\in Ker\, B} \frac{\displaystyle\int_\Gamma (\mathbf{v}\cdot\mathbf{n})^2}{\displaystyle\int_\Omega |\boldsymbol{\varepsilon}(\mathbf{v})|^2} > 0 \,, \quad (29)$$

we see that condition (28) still holds whenever

$$\gamma < \frac{2\mu}{\alpha_M} \,. \quad (30)$$

We now give an estimate of α_M by means of the following Proposition, whose proof is reported in the Appendix Section A.1 (see also Auricchio et al., 2005b).

Proposition 3.2. *Suppose that* $\Omega = (-1,1) \times (-1,1)$*. Then* $\alpha_M \leq 2/3$.

Remark 3.3. We remark that the above estimate of α_M may not be sharp. However, it is sufficient for our subsequent considerations. Moreover, we notice that condition (28) cannot hold for arbitrarily large values of γ. Indeed, let $\mathbf{w}$ be a function in $KerB$ which does not vanish on Γ. Hence

$$\int_\Gamma (\mathbf{w} \cdot \mathbf{n})^2 > 0.$$

Choosing γ^* as

$$\gamma^* := \frac{2\mu \displaystyle\int_\Omega |\boldsymbol{\varepsilon}(\mathbf{w})|^2}{\displaystyle\int_\Gamma (\mathbf{w} \cdot \mathbf{n})^2}, \tag{31}$$

it follows that

$$2\mu \int_\Omega |\boldsymbol{\varepsilon}(\mathbf{w})|^2 - \gamma \int_\Gamma (\mathbf{w} \cdot \mathbf{n})^2 \leq 0 \tag{32}$$

for all $\gamma \geq \gamma^*$. Therefore, coercivity on the kernel surely fails for sufficiently large values of γ.

A consequence of Proposition 3.2 is that the linearized continuous problem (16) is well-posed and positive-definite on the relevant kernel $Ker\,B$ if

$$\gamma \in \left(-\infty, \frac{2\mu}{\alpha_M} \right) \supseteq (-\infty, 3\mu) \ . \tag{33}$$

4 Discrete Stability Range: Some Theoretical Results

For a finite analysis of the problem presented in Section 3, one should obviously focus on schemes which are reliable at least in the infinitesimal strain regime, otherwise even the first iteration step will fail. This can be accomplished by considering finite element methods satisfying the discrete *inf-sup* condition, as the ones studied in this Section.

As proved in Subsection 3.1, the continuous linearized problems (16) are surely stable, in the sense that both the *inf-sup* and the *coercivity on the kernel* conditions hold true, whenever the parameter γ satisfies

$$\gamma \in (-\infty, 3\mu) \ . \tag{34}$$

On the other hand, the linearized problems (16) become unstable for suitable large values of γ (cf. Remark 3.3). We point out that this lack of stability is only addressed to a failure of the *coercivity on the kernel* condition. Indeed, for the proposed problem the *inf-sup* condition is *independent* of γ, and therefore it is the same one required for the small deformation framework.

Since a reliable numerical approximation should be able to correctly reproduce the stability properties of the continuous problem, the discussion above highlights the importance of studying whether or not a given finite element method satisfies a discrete *coercivity on the kernel* condition, at least for γ in the range shown by (34). With this respect, we consider the MINI element (cf. Arnold et al., 1984) and the QME element (cf. Pantuso and Bathe, 1995), rigorously proving that their stability range are somehow quite different from the continuous problem one. However, we point out that our theoretical analysis is far from being *complete*, although in accordance with the numerical tests presented in Section 5.

4.1 The MINI Element

We now consider the discretized counterpart of problem (16), using the MINI element (cf. Arnold et al., 1984).

Let $\mathcal{T}_h$ be a triangular mesh of Ω, h being the meshsize. For the discretization of the displacement field, we take

$$U_h = \left\{ \mathbf{v}_h \in U \ : \ \mathbf{v}_{h|T} \in \mathcal{P}_1(T)^2 + \mathcal{B}(T)^2 \qquad \forall T \in \mathcal{T}_h \right\} , \tag{35}$$

where $\mathcal{P}_1(T)$ is the space of linear functions on T, and $\mathcal{B}(T)$ is the linear space generated by b_T, the standard cubic bubble function on T. For the pressure discretization, we take

$$P_h = \left\{ q_h \in H^1(\Omega) \ : \ q_{h|T} \in \mathcal{P}_1(T) \qquad \forall T \in \mathcal{T}_h \right\} . \tag{36}$$

Therefore, the discretization of problem (16) reads as follows.

Find $(\mathbf{u}_h; p_h) \in U_h \times P_h$ such that:

$$\begin{cases} 2\mu \displaystyle\int_\Omega \boldsymbol{\varepsilon}(\mathbf{u}_h) : \boldsymbol{\varepsilon}(\mathbf{v}_h) - \gamma \int_\Omega r \left(\nabla \mathbf{u}_h\right)^T : \nabla \mathbf{v}_h \\ \qquad\qquad + \displaystyle\int_\Omega p_h \, \mathrm{div} \mathbf{v}_h = \Delta\gamma \int_\Omega \mathbf{f} \cdot \mathbf{v}_h \qquad \forall \mathbf{v}_h \in U_h \\ \displaystyle\int_\Omega q_h \mathrm{div} \mathbf{u}_h = 0 \qquad\qquad \forall q_h \in P_h \, . \end{cases} \tag{37}$$

Introducing the discrete kernel as

$$K_h = \left\{ \mathbf{v}_h \in U_h \ : \ \int_\Omega q_h \mathrm{div} \mathbf{u}_h = 0 \qquad \forall q_h \in P_h \right\} , \tag{38}$$

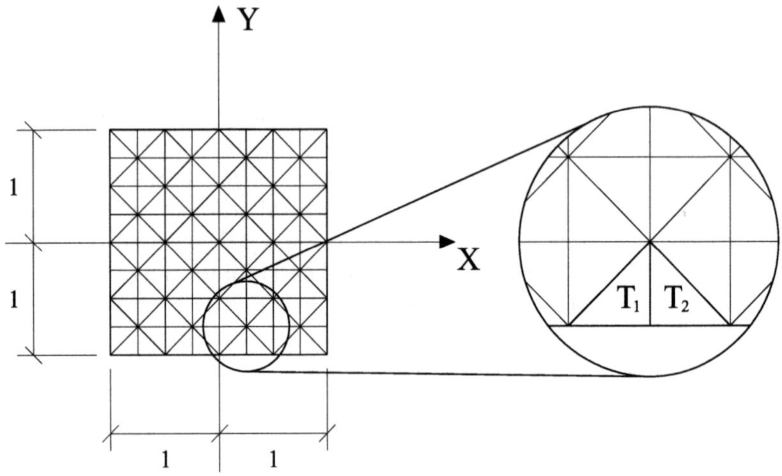

Figure 2. Triangular mesh for the MINI element.

we are interested in analyzing for which γ there exists a constant $c_M(\gamma, \mu) > 0$ such that

$$a_\gamma(\nabla \mathbf{v}_h, \nabla \mathbf{v}_h) \geq c_M(\gamma, \mu) \int_\Omega |\, \boldsymbol{\varepsilon}(\mathbf{v}_h)\,|^2 \qquad \forall \mathbf{v}_h \in K_h \,, \tag{39}$$

where

$$a_\gamma(\nabla \mathbf{v}_h, \nabla \mathbf{v}_h) := 2\mu \int_\Omega \boldsymbol{\varepsilon}(\mathbf{v}_h) : \boldsymbol{\varepsilon}(\mathbf{v}_h) - \gamma \int_\Omega r\,(\nabla \mathbf{v}_h)^T : \nabla \mathbf{v}_h \,. \tag{40}$$

We consider the meshes used for the numerical tests presented in Section 5. Therefore $\mathcal{T}_h$ is built by squares of side $2h$, each divided into eight triangles according with the pattern shown in Fig. 2.

We have the following results.

Proposition 4.1. *For problem (37), the discrete coercivity on the kernel condition (39) does not hold, whenever* $\gamma > 3\mu/2$.

Proposition 4.2. *For problem (37), the discrete coercivity on the kernel condition (39) is satisfied, independently of the meshsize, whenever* $\gamma < \mu$.

For the proofs of these Propositions see Auricchio et al. (2005b) or Appendix Sections A.2 and A.3.

4.2 The QME Element

We now consider the discretized counterpart of problem (16), using the QME quadrilateral method proposed by Pantuso and Bathe (1995), based on the *Enhanced Strain Technique* (cf. Simo and Rifai, 1990, for instance). This scheme optimally performs in small deformation regimes, as theoretically proved by Lovadina (1997).

For our analysis we consider uniform meshes $\mathcal{T}_h$ formed by equal square elements K with side length h, as the ones used in the numerical tests of Section 5.

The Pantuso-Bathe element is described by the following choice of spaces. For the discretization of the displacement field, we take

$$U_h = \left\{ \mathbf{v}_h \in U \ : \ \mathbf{v}_{h|K} \in \mathcal{Q}_1(K)^2 \qquad \forall K \in \mathcal{T}_h \right\} , \tag{41}$$

where $\mathcal{Q}_1(K)$ is the standard space of bilinear functions. For the pressure discretization, we take

$$P_h = \left\{ q_h \in H^1(\Omega) \ : \ q_{h|K} \in \mathcal{Q}_1(K) \qquad \forall K \in \mathcal{T}_h \right\} . \tag{42}$$

Furthermore, the *Enhanced Strain* space is described by

$$S_h = \left\{ \mathbf{E}_h \in (L^2(\Omega))^4 \ : \ \mathbf{E}_{h|K} \in E_6(K) \qquad \forall K \in \mathcal{T}_h \right\} . \tag{43}$$

Above, $E_6(K)$ is the space of tensor-valued functions defined on K, spanned by the following shape functions

$$\begin{bmatrix} \alpha_1\xi + \alpha_2\xi\eta & ; & \alpha_3\xi \\ \alpha_4\eta & ; & \alpha_5\eta + \alpha_6\xi\eta \end{bmatrix} \qquad \text{with } \alpha_i \in \mathbf{R} , \tag{44}$$

where (ξ, η) denotes the standard local coordinates on K.

Therefore, the discretization of problem (16) reads as follows.

Find $(\mathbf{u}_h, \mathbf{H}_h; p_h) \in (U_h \times S_h) \times P_h$ such that:

$$\begin{cases} 2\mu \int_\Omega (\nabla \mathbf{u}_h + \mathbf{H}_h)^S : (\nabla \mathbf{v}_h + \mathbf{E}_h)^S - \\ \qquad \gamma \int_\Omega r\, (\nabla \mathbf{u}_h + \mathbf{H}_h)^T : (\nabla \mathbf{v}_h + \mathbf{E}_h) \\ \\ + \int_\Omega p_h \, (\mathrm{div}\mathbf{v}_h + \mathrm{tr}\mathbf{E}_h) = \Delta\gamma \int_\Omega \mathbf{f} \cdot \mathbf{v}_h \qquad \forall (\mathbf{v}_h, \mathbf{E}_h) \in U_h \times S_h \\ \\ \int_\Omega q_h (\mathrm{div}\mathbf{u}_h + \mathrm{tr}\mathbf{H}_h) = 0 \qquad\qquad \forall q_h \in P_h \ . \end{cases} \tag{45}$$

Above and in the sequel, we denote with "tr" the trace operator acting on tensors.

Introducing the discrete kernel as

$$K_h = \left\{ (\mathbf{v}_h, \mathbf{E}_h) \in U_h \times S_h \ : \ \int_\Omega q_h (\mathrm{div}\mathbf{v}_h + \mathrm{tr}\mathbf{E}_h) = 0 \qquad \forall q_h \in P_h \right\} , \tag{46}$$

we are interested in analyzing for which γ there exists a constant $c_E(\gamma, \mu)$ such that (cf. (18))

$$a_\gamma(\nabla \mathbf{v}_h + \mathbf{E}_h, \nabla \mathbf{v}_h + \mathbf{E}_h) \geq c_E(\gamma, \mu) \int_\Omega \left(|\boldsymbol{\varepsilon}(\mathbf{v}_h)|^2 + |\mathbf{E}_h|^2 \right) \tag{47}$$

for every $(\mathbf{v}_h, \mathbf{E}_h) \in K_h$. We have the following result (see Auricchio et al., 2005b).

Proposition 4.3. *For the choice (41)–(44), the discrete coercivity on the kernel condition does not hold, whenever $\gamma > \mu$.*

The proof for this Proposition is reported in the Appendix Section A.4.

5 Numerical Tests

We now study the computational performance of some specific finite element interpolations on the model problem presented in previous Sections. In particular, we wish to numerically detect the stability range of the elements under investigation, and, when it is possible, to compare such numerical results with the theoretical ones previously obtained for some *discrete* problems as well as for the *continuous* one.

The interpolation schemes we deal with in this Section are

- MINI – Triangular element with piecewise linear continuous approximation for both displacements and pressure, with the displacements enriched by a cubic bubble as proposed by Arnold et al. (1984);
- QME – Quadrilateral element with piecewise bilinear continuous approximation for both displacements and pressure, enriched by the Enhanced Strains as proposed by Pantuso and Bathe (1995);
- Q2P1 – Quadrilateral element with piecewise biquadratic continuous approximation for displacements and piecewise linear approximation for pressure;
- QE1 – Quadrilateral element with piecewise bilinear continuous approximation for displacements, enriched by the Enhanced Strains as proposed by Simo and Armero (1992);

- QE2 – Quadrilateral element with piecewise bilinear continuous approximation for displacements, enriched by the Enhanced Strains as proposed by Glaser and Armero (1997).

All the schemes have been implemented within the research-oriented finite element environment FEAP (see Taylor, 2001).

The model problem on which we perform our numerical studies is the square incompressible block clamped on three edges and subjected to a vertical uniform body load depicted in Fig. 1 (cf. Section 3). We remark that for non-mixed formulations only nearly incompressible conditions can be considered.

Assuming to express respectively forces and lengths in KN and m, we set $\mu = 40$ and we select a body load $\gamma(0,1)^T$, where γ plays the role of load multiplier; moreover for non-mixed formulations we set $\lambda = 4 \cdot 10^5$ in order to make the problem nearly incompressible.

Here and in the following, we find convenient to express the numerical results in terms of the nondimensional quantity $\tilde{\gamma}$ defined as $\tilde{\gamma} = \gamma L/\mu$ with L some problem characteristic length, set equal to 1 for simplicity and consistently with the model problem.

For a given interpolation scheme and for a given mesh, in order to numerically detect the element stability range we progressively increase the load multiplier γ, adopting an iterative Newton's scheme to obtain the solution corresponding to the single load value from the solution corresponding to the previous load value. In particular, we increase the load multiplier until some form of numerical instability appears (e.g., appearance of negative eigenvalues, divergence of the solution, etc.); we indicate the load multiplier corresponding to the first appearance of numerical instabilities as γ_{cr} and the corresponding nondimensional multiplier as $\tilde{\gamma}_{cr} = \gamma_{cr} L/\mu$. To investigate very large load multiplier intervals, we adopt different increments $\Delta\gamma$ depending on the load level (cf. Table 1). Accordingly, we do a series of sequential analyses. In the first one, we increase the load multiplier γ from 0 to a maximum value $\gamma = 10^2$ using load increments $\Delta\gamma = 10^{-1}$. If no instability is detected, we repeat the analysis increasing now the load multiplier γ from 0 to $\gamma = 10^3$ using load increments $\Delta\gamma = 1$, and we keep on going until a numerical instability appears.

Clearly, the analyses are performed starting from $\tilde{\gamma} = 0$ for both positive and negative loading conditions, i.e. for $\tilde{\gamma} < 0$ and $\tilde{\gamma} > 0$. If we do not detect numerical instabilities even for extremely large values of the load multiplier, we set $\tilde{\gamma}_{cr} = \infty$.

Finally, to make immediately clear the comparisons among different elements and meshes, we identify any given mesh with the label "$n \times n$", being n the number of displacement nodes per side of the mesh.

Table 1. Load increments $\Delta\gamma$ (depending on the load level γ) for the Newton's scheme.

$\Delta\gamma = 10^{-1}$	up to	$\gamma = 10^2$
$\Delta\gamma = 1$	up to	$\gamma = 10^3$
$\Delta\gamma = 10$	up to	$\gamma = 10^4$
$\Delta\gamma = 10^2$	up to	$\gamma = 10^5$
$\Delta\gamma = 10^3$	up to	$\gamma = 10^6$

Table 2 reports the stability limits for the different considered interpolation schemes when applied to our model problem. From such results the following observations arise.

- The theoretical predictions for the MINI interpolation scheme are that the *discrete* problem is stable for $-\infty < \tilde{\gamma} < 1$ and unstable for $\tilde{\gamma} > 3/2$ (cf. Propositions 4.1 and 4.2). The results presented in Table 2 show that the corresponding *numerical* problem is stable for $-\infty < \tilde{\gamma} < 1$, unstable for $\tilde{\gamma} > 3/2$. In particular, the stability upper limit approaches monotonically 1 from above during the mesh refinement (i.e. for $h \to 0$). Accordingly, for the MINI interpolation scheme the numerical results confirm the theoretical predictions.
- The theoretical predictions for the QME interpolation scheme are that for sufficiently small h the *discrete* problem is unstable for $\tilde{\gamma} > 1$ (cf. Proposition 4.3). The numerical experiments seem to indicate that for sufficiently small h the corresponding *numerical* problem is stable for $-\infty < \tilde{\gamma} < 1$ and unstable for $\tilde{\gamma} > 1$. In particular, the stability lower limit is decreasing almost linearly with h, approaching $-\infty$ for $h \to 0$, while the stability upper limit is about 1 but it does not seem to have a smooth convergence. Accordingly, also for the QME interpolation scheme the numerical results confirm the theoretical predictions.
- We do not have theoretical prediction for the Q2P1 *discrete* problem, except the trivial one telling that the problem is stable in the range $-1 < \tilde{\gamma} < 1$ (cf. the beginning of Proposition 4.2 proof). However, the results relative to the Q2P1 interpolation scheme indicate that the *numerical* problem is stable for $-\infty < \tilde{\gamma} < 1$.
- Also for the enhanced *discrete* problems (elements QE1 and QE2), only the trivial theoretical prediction is available (stability for $-1 < \tilde{\gamma} < 1$). For element QE1, it seems that the trivial range is the one recovered from the numerical results, as h becomes small; element QE2 has a similar behaviour for positive values of $\tilde{\gamma}$, but the stability range in compression appears to be much larger than with QE1, even if it

seems to get smaller as h decreases (but a smooth convergence to a stable value cannot be observed).
We remark that in Table 2, where the numerical results referred to elements QE1 and QE2 are reported, some values are followed by one or two asterisks. Their meaning is that no instabilities are detected when using in the analysis a load level of $\gamma = 10^2$, but anyway an instability appears for $\gamma < 10^2$ when an higher load level is considered (along with its corresponding load increment, cf. Table 1). In particular, one asterisk (*) indicates that a load level of $\gamma = 10^3$ is used, while two asterisks (**) indicate that a load level of $\gamma = 10^4$ is used.

Table 2. Stability limits for the studied numerical schemes on the model problem of Fig. 1 in the cases of compression ($\tilde{\gamma}_{cr}^-$) and tension ($\tilde{\gamma}_{cr}^+$). The asterisks mean that (nondimensional) load levels of $\tilde{\gamma} = \pm 25$ (*) or $\tilde{\gamma} = \pm 250$ (**) are used instead of $\tilde{\gamma} = \pm 2.5$

	MINI		QME		Q2P1	
mesh	$\tilde{\gamma}_{cr}^-$	$\tilde{\gamma}_{cr}^+$	$\tilde{\gamma}_{cr}^-$	$\tilde{\gamma}_{cr}^+$	$\tilde{\gamma}_{cr}^-$	$\tilde{\gamma}_{cr}^+$
5×5	$-\infty$	1.28	-52.8	1.21	$-\infty$	1.46
9×9	$-\infty$	1.22	-150	1.07	$-\infty$	1.16
17×17	$-\infty$	1.19	-335	0.998	$-\infty$	1.07
33×33	$-\infty$	1.18	-723	0.978	$-\infty$	1.03
65×65	$-\infty$	1.17	-1480	0.980	$-\infty$	1.01

	QE1		QE2	
mesh	$\tilde{\gamma}_{cr}^-$	$\tilde{\gamma}_{cr}^+$	$\tilde{\gamma}_{cr}^-$	$\tilde{\gamma}_{cr}^+$
5×5	-2.38	2.38	-1060	4.53
9×9	-1.75**	1.60**	-723	1.88*
17×17	-1.50**	6.80	-588	1.43*
33×33	-1.87	1.82	-460	1.25*
65×65	-1.65	1.10	-645	1.13

As a conclusion to our study, we recall that the *continuous* problem is stable at least for $-\infty < \tilde{\gamma} < 3$ (see (33)), and that for a sufficiently large value of $\tilde{\gamma}$ the problem gets unstable (see Remark 3.3). Therefore, we may state that all the interpolation schemes fail in properly detecting the stability range of the continuous problem. However we have to notice that all the considered mixed formulations show a much better behaviour in compression with respect to the (non-mixed) enhanced ones. Finally, the MINI and the Q2P1 interpolation schemes seem to be the most effective. Indeed, they are the elements able to reproduce more closely (but still in a deficient form) the continuous problem behavior, as shown in Table 3, where we qualitatively evaluate the performance of the studied numerical schemes. The meaning of the symbols used in the Table is the following:

$$\begin{aligned} \checkmark\checkmark &: \text{optimal performance;} \\ \checkmark &: \text{good performance;} \\ \spadesuit &: \text{poor performance;} \\ \spadesuit\spadesuit &: \text{very poor performance.} \end{aligned}$$

Table 3. Qualitative performance evaluation of the studied numerical schemes.

MINI		QME		Q2P1	
$\tilde{\gamma} < 0$	$\tilde{\gamma} > 0$	$\tilde{\gamma} < 0$	$\tilde{\gamma} > 0$	$\tilde{\gamma} < 0$	$\tilde{\gamma} > 0$
$\checkmark\checkmark$	$\spadesuit$	$\checkmark$	$\spadesuit$	$\checkmark\checkmark$	$\spadesuit$

QE1		QE2	
$\tilde{\gamma} < 0$	$\tilde{\gamma} > 0$	$\tilde{\gamma} < 0$	$\tilde{\gamma} > 0$
$\spadesuit\spadesuit$	$\spadesuit$	$\spadesuit$	$\spadesuit$

6 Conclusions

We have proposed a simple 2D finite-strain problem depending on a loading parameter, for which a trivial solution can be easily computed. We have proved the stability of such a solution whenever the loading parameter stays in a suitable range of values. Furthermore, we have considered and analyzed the problem discretization by means of some finite element schemes, which are known to efficiently behave in the framework of small deformation problems. In particular, we have proved that the elements fail to properly detect the problem stability range. We have also presented several numerical experiments, in accordance with the theoretical predictions, showing which elements behave better on the considered model problem. However, we notice that a completely satisfactory analysis of finite element schemes for large deformation problems is still missing. Therefore, we think it could be valuable to propose other benchmarks in the spirit of the present work.

Bibliography

F. Armero. On the locking and stability of finite elements in finite deformation plane strain problems. *Computers & Structures*, 75:261–290, 2000.

D. N. Arnold, F. Brezzi, and M. Fortin. A stable finite element for the stokes equation. *Calcolo*, 21:337–344, 1984.

F. Auricchio, L. Beirão da Veiga, C. Lovadina, and A. Reali. An analysis of some mixed-enhanced finite element for plane linear elasticity. *Computer Methods in Applied Mechanics and Engineering*, 194:2947–2968, 2005a.

F. Auricchio, L. Beirão da Veiga, C. Lovadina, and A. Reali. A stability study of some mixed finite elements for large deformation elasticity problems. *Computer Methods in Applied Mechanics and Engineering*, 194:1075–1092, 2005b.

K. J. Bathe. *Finite Element Procedures.* Prentice Hall, Englewood Cliffs, NJ, 1996.

J. Bonet and R. D. Wood. *Nonlinear Continuum Mechanics for Finite Element Analysis.* Cambridge University Press, 1997.

D. Braess. Enhanced assumed strain elements and locking in membrane problems. *Computer Methods in Applied Mechanics and Engineering*, 165:155–174, 1998.

F. Brezzi and M. Fortin. *Mixed and Hybrid Finite Element Methods.* Springer-Verlag, New York, 1991.

P. G. Ciarlet. *The Finite Element Methods for Elliptic Problems.* North-Holland, Amsterdam, 1978.

S. Glaser and F. Armero. On the formulation of enhanced strain finite elements in finite deformations. *Engineering Computations*, 14:759–791, 1997.

T. J. R. Hughes. *The Finite Element Method: Linear Static and Dynamic Finite Element Analysis.* Dover Publications, Mineola, New York, 2000.

O. Klaas, A. M. Maniatty, and M. S. Shephard. A stabilized mixed petrov-galerkin finite element method for finite elasticity. formulation for linear displacement and pressure interpolation. *Computer Methods in Applied Mechanics and Engineering*, 180:65–79, 1999.

P. Le Tallec. Existence and approximation results for nonlinear mixed problems: application to incompressible finite elasticity. *Numerische Mathematik*, 38:365–382, 1982.

P. Le Tallec. Numerical methods for nonlinear three-dimensional elasticity. In P. G. Ciarlet and J. L. Lions, editors, *Handbook of Numerical Analysis*, volume III, pages 465–622. Elsevier Science, North Holland, 1994.

C. Lovadina. Analysis of strain-pressure finite element methods for the stokes problem. *Numerical Methods for PDE's*, 13:717–730, 1997.

C. Lovadina and F. Auricchio. On the enhanced strain technique for elasticity problems. *Computers & Structures*, 81:777–787, 2003.

A. M. Maniatty, Y. Liu, O. Klaas, and M. S. Shephard. Higher order stabilized finite element method for hyperelastic finite deformation. *Computer Methods in Applied Mechanics and Engineering*, 191:1491–1503, 2002.

J. C. Nagtegaal and D. D. Fox. Using assumed enhanced strain elements for large compressive deformation. *International Journal of Solids and Structures*, 33:3151–3159, 1996.

D. Pantuso and K. J. Bathe. A four-node quadrilateral mixed-interpolated element for solids and fluids. *Mathematical Models & Methods in Applied Sciences*, 5:1113–1128, 1995.

D. Pantuso and K. J. Bathe. On the stability of mixed finite elements in large strain analysis of incompressible solids. *Finite Elements in Analysis and Design*, 28:83–104, 1997.

B. D. Reddy and J. C. Simo. Stability and convergence of a class of enhanced strain methods. *SIAM Journal on Numerical Analysis*, 32:705–1728, 1995.

S. Reese, M. Küssner, and B. D. Reddy. A new stabilization technique for finite elements in non-linear elasticity. *International Journal for Numerical Methods in Engineering*, 44:1617–1652, 1999.

S. Reese and P. Wriggers. A stabilization technique to avoid hourglassing in finite elasticity. *International Journal for Numerical Methods in Engineering*, 48:79–109, 2000.

J. C. Simo and F. Armero. Geometrically nonlinear enhanced strain mixed methods and the method of incompatible modes. *International Journal for Numerical Methods in Engineering*, 33:1413–1449, 1992.

J. C. Simo and M. S. Rifai. A class of mixed assumed strain methods and the method of incompatible modes. *International Journal for Numerical Methods in Engineering*, 29:1595–1638, 1990.

R. L. Taylor. FEAP: A Finite Element Analysis Program, programmer manual. http://www.ce.berkeley.edu/~rlt/feap/, 2001.

P. Wriggers and S. Reese. A note on enhanced strain methods for large deformations. *Computer Methods in Applied Mechanics and Engineering*, 135:201–209, 1996.

A Mathematical Proofs

A.1 Proof for Proposition 3.2

Take $\mathbf{v} = (v_1, v_2) \in (Ker\, B) \cap C^1(\bar{\Omega})$ and observe that on Γ we have $\mathbf{v} \cdot \mathbf{n} = v_2$. Therefore

$$\int_\Gamma (\mathbf{v} \cdot \mathbf{n})^2 = \int_\Gamma |v_2|^2. \tag{48}$$

Since for every $X \in [-1, 1]$ it holds $v_2(X, -1) = 0$, we get

$$v_2(X, 1) = \int_{-1}^{1} v_{2,Y}(X, Y)dY \ .$$

Hence, the Cauchy-Schwarz inequality gives

$$\int_\Gamma |v_2|^2 \leq 2 \int_\Omega |v_{2,Y}|^2 \ . \tag{49}$$

Using $\text{div}\mathbf{v} = 0$, we obtain

$$\int_\Gamma |v_2|^2 \leq \int_\Omega |v_{1,X}|^2 + \int_\Omega |v_{2,Y}|^2 \leq \int_\Omega |\ \varepsilon_{11}(\mathbf{v})\ |^2 + \int_\Omega |\ \varepsilon_{22}(\mathbf{v})\ |^2 \tag{50}$$

for all $\mathbf{v} \in (Ker\, B) \cap C^1(\bar{\Omega})$.

Before proceeding, we need to introduce some notation. For every point $P = (X, 1)$ with $X \in [-1, 1]$, we define the oriented rectilinear path from point P_X^1 to point P (see Fig. 3), parameterized by

$$\gamma_X^1(s) = (-1 + s/\sqrt{2}, -X + s/\sqrt{2}) \qquad s \in [0, \sqrt{2}(X + 1)] \ , \tag{51}$$

and with unit tangent vector

$$\boldsymbol{\tau}_1 = (1/\sqrt{2}, 1/\sqrt{2}) \ . \tag{52}$$

Similarly, for every $X \in [-1, 1]$, we define the oriented rectilinear path from point P_X^2 to point P (see Fig. 3), parameterized by

$$\gamma_X^2(s) = (1 - s/\sqrt{2}, X + s/\sqrt{2}) \qquad s \in [0, \sqrt{2}(1 - X)] \ , \tag{53}$$

and with unit tangent vector

$$\boldsymbol{\tau}_2 = (-1/\sqrt{2}, 1/\sqrt{2}) \ . \tag{54}$$

Moreover, we introduce the union path $\gamma_X = \gamma_X^1 \cup \{-\gamma_X^2\}$ going from point P_X^1 to point P_X^2.

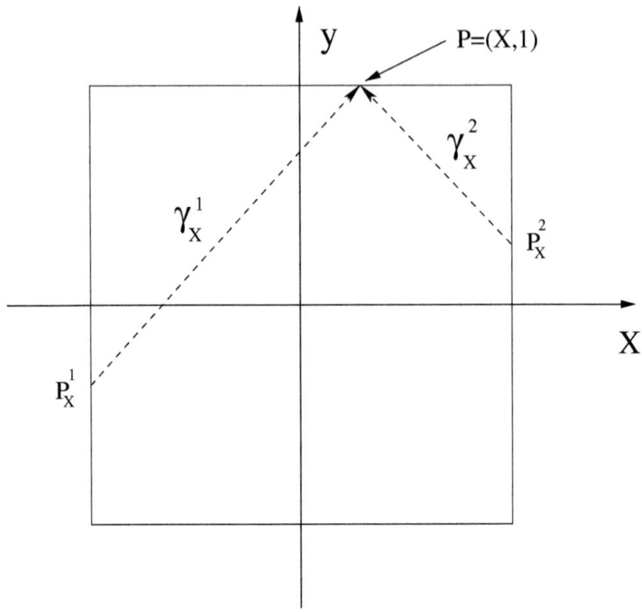

Figure 3. Oriented paths γ_X^1 and γ_X^2 on Ω.

For all $X \in [-1,1]$, it clearly holds

$$v_2(X,1) = \frac{1}{\sqrt{2}}\mathbf{v}(X,1)\cdot\boldsymbol{\tau}_1 + \frac{1}{\sqrt{2}}\mathbf{v}(X,1)\cdot\boldsymbol{\tau}_2 \ . \tag{55}$$

Observing that $\mathbf{v}$ vanishes on $\{-1\}\times[-1,1]$ and integrating along γ_X^1, the first term in the right-hand side of (55) can be written as

$$\mathbf{v}(X,1)\cdot\boldsymbol{\tau}_1 = \int_0^{\sqrt{2}(X+1)} \nabla(\mathbf{v}\cdot\boldsymbol{\tau}_1)(\gamma_X^1(s))\cdot\boldsymbol{\tau}_1 \,\mathrm{d}s \ , \tag{56}$$

which, after some simple algebra, gives

$$\mathbf{v}(X,1)\cdot\boldsymbol{\tau}_1 = \int_0^{\sqrt{2}(X+1)} \boldsymbol{\varepsilon}\big(\mathbf{v}(\gamma_X^1(s))\big)\boldsymbol{\tau}_1\cdot\boldsymbol{\tau}_1 \,\mathrm{d}s \ . \tag{57}$$

Recalling that $\mathrm{div}\mathbf{v} = 0$, it is easy to see that in Ω we have

$$\boldsymbol{\varepsilon}(\mathbf{v})\boldsymbol{\tau}_1\cdot\boldsymbol{\tau}_1 = \varepsilon_{12}(\mathbf{v}) \ . \tag{58}$$

Using (58) and (57), we obtain

$$\mathbf{v}(X,1)\cdot\boldsymbol{\tau}_1 = \int_0^{\sqrt{2}(X+1)} \varepsilon_{12}(\boldsymbol{\gamma}^1_X(s))\,\mathrm{d}s\ , \tag{59}$$

or, in more compact form,

$$\mathbf{v}(X,1)\cdot\boldsymbol{\tau}_1 = \int_{\gamma^1_X} \varepsilon_{12}(\mathbf{v})\ . \tag{60}$$

Treating similarly the second term in the right-hand side of (55), but using the path $\boldsymbol{\gamma}^2_X$, it can be shown that

$$\mathbf{v}(X,1)\cdot\boldsymbol{\tau}_2 = -\int_{\gamma^2_X} \varepsilon_{12}(\mathbf{v})\ . \tag{61}$$

Recalling (55), from (60) (61) we get

$$v_2(X,1) = \frac{1}{\sqrt{2}}\Big(\int_{\gamma^1_X}\varepsilon_{12}(\mathbf{v}) - \int_{\gamma^2_X}\varepsilon_{12}(\mathbf{v})\Big) = \frac{1}{\sqrt{2}}\int_{\gamma_X}\varepsilon_{12}(\mathbf{v})\ , \tag{62}$$

so that we obtain

$$\int_\Gamma |\, v_2\,|^2 = \int_{-1}^1 |\, v_2(X,1)\,|^2\ \mathrm{d}X = \frac{1}{2}\int_{-1}^1 \left|\int_{\gamma_X}\varepsilon_{12}(\mathbf{v})\right|^2 \mathrm{d}X\ . \tag{63}$$

From (63) and observing that $|\,\gamma^1_X\,| + |\,\gamma^2_X\,| = 2\sqrt{2}$ for all $X \in [-1,1]$, the Cauchy-Schwarz inequality yields

$$\int_\Gamma |\, v_2\,|^2 \le \sqrt{2}\int_{-1}^1 \Big(\int_{\gamma_X} |\,\varepsilon_{12}(\mathbf{v})\,|^2\Big)\,\mathrm{d}X\ . \tag{64}$$

Setting

$$\Omega_1 = \big\{(X,Y)\in\Omega\ :\ Y \ge X\big\} \qquad \Omega_2 = \big\{(X,Y)\in\Omega\ :\ Y \ge -X\big\}\ , \tag{65}$$

we easily get, using a simple change of variables,

$$\int_{-1}^1\int_{\gamma^1_X} |\,\varepsilon_{12}(\mathbf{v})\,|^2 = \sqrt{2}\int_{\Omega_1} |\,\varepsilon_{12}(\mathbf{v})\,|^2 \le \sqrt{2}\int_\Omega |\,\varepsilon_{12}(\mathbf{v})\,|^2\ , \tag{66}$$

and

$$\int_{-1}^1\int_{\gamma^2_X} |\,\varepsilon_{12}(\mathbf{v})\,|^2 = \sqrt{2}\int_{\Omega_2} |\,\varepsilon_{12}(\mathbf{v})\,|^2 \le \sqrt{2}\int_\Omega |\,\varepsilon_{12}(\mathbf{v})\,|^2\ . \tag{67}$$

From (64), splitting the integral on γ_X into the contributions on γ_X^1 and γ_X^2, using bounds (66)-(67) we obtain

$$\int_\Gamma |\, v_2 \,|^2 \leq 4 \int_\Omega |\, \varepsilon_{12}(\mathbf{v}) \,|^2 \ . \tag{68}$$

Recalling (48), from (50) and (68) it follows

$$\int_\Gamma |\mathbf{v} \cdot \mathbf{n}|^2 \leq \frac{2}{3} \int_\Omega |\, \boldsymbol{\varepsilon}(\mathbf{v}) \,|^2 \tag{69}$$

for all $\mathbf{v} \in (Ker\, B) \cap C^1(\bar{\Omega})$.

By density the same estimate holds in $Ker\, B$, so that

$$\alpha_M = \sup_{\mathbf{v} \in Ker\, B} \frac{\displaystyle\int_\Gamma (\mathbf{v} \cdot \mathbf{n})^2}{\displaystyle\int_\Omega |\boldsymbol{\varepsilon}(\mathbf{v})|^2} \leq \frac{2}{3} \ , \tag{70}$$

which concludes the proof. □

A.2 Proof for Proposition 4.1

Take any two adjacent triangles T_1 and T_2, both with a side contained in $[-1, 1] \times \{-1\}$ (see Fig. 2). Consider the function $\mathbf{w}_h = (w_1, w_2) \in U_h$ defined as

$$w_1 = 0 \qquad ; \qquad w_2 = \begin{cases} b_{T_1} & \text{on } T_1 \\ -b_{T_2} & \text{on } T_2 \\ 0 & \text{otherwise} \ . \end{cases} \tag{71}$$

It is easy to see that

$$\int_\Omega q_h \mathrm{div} \mathbf{w}_h = \int_{T_1 \cup T_2} q_h \mathrm{div} \mathbf{w}_h = - \int_{T_1 \cup T_2} \nabla q_h \cdot \mathbf{w}_h = 0 \qquad \forall q_h \in P_h \ , \tag{72}$$

so that $\mathbf{w}_h \in K_h$. From (40) and (71), we get

$$a_\gamma(\mathbf{w}_h, \mathbf{w}_h) = 2\mu \int_{T_1 \cup T_2} \left[(w_{2,Y})^2 + \frac{1}{2} (w_{2,X})^2 \right] - \gamma \int_{T_1 \cup T_2} (1 - Y)\, (w_{2,Y})^2 \ . \tag{73}$$

Using that for our particular mesh it holds

$$\int_T (b_{T,Y})^2 = \int_T (b_{T,X})^2 \qquad \forall T \in \mathcal{T}_h \ , \tag{74}$$

and noting that on $T_1 \cup T_2$ we obviously have $Y < h-1$, from (71) and (73) it follows

$$a_\gamma(\mathbf{w},\mathbf{w}) \leq (3\mu - (2-h)\gamma) \int_{T_1\cup T_2} (w_{2,Y})^2 \ . \tag{75}$$

Therefore the form $a_\gamma(\cdot,\cdot)$ fails to be coercive on K_h whenever γ and h are such that

$$3\mu - (2-h)\gamma \leq 0 \ . \tag{76}$$

It follows that, for $\gamma > 3\mu/2$, the coercivity on the kernel breaks down, provided h is sufficiently small. □

A.3 Proof for Proposition 4.2

It is easy to check that, for $-\mu < \gamma < \mu$ and $-1 < Y < 1$, there exists a constant $c > 0$ such that

$$2\mu \mathbf{A}^S : \mathbf{A}^S - \gamma r \mathbf{A}^T : \mathbf{A} \geq c \mid \mathbf{A}^S \mid^2 \tag{77}$$

for every second order tensors $\mathbf{A}$. From the pointwise positivity property (77) it easily follows that, for $-\mu < \gamma < \mu$, the bilinear form (see (40))

$$a_\gamma(\nabla\mathbf{u},\nabla\mathbf{v}) = 2\mu \int_\Omega \boldsymbol{\varepsilon}(\mathbf{u}) : \boldsymbol{\varepsilon}(\mathbf{v}) - \gamma \int_\Omega r\,(\nabla\mathbf{u})^T : \nabla\mathbf{v} \tag{78}$$

is coercive on the whole space U (and therefore in particular for $K_h \subseteq U_h \subseteq U$).

We will now prove that for $\mathbf{v}_h \subset K_h$ it holds

$$\int_\Omega r\,(\nabla\mathbf{v}_h)^T : \nabla\mathbf{v}_h \geq 0 \ . \tag{79}$$

Once estimate (79) has been established, also for $\gamma < 0$ the coercivity property (39) immediately follows, and the proof is complete.

We first observe that

$$(\nabla\mathbf{v}_h)^T = (\mathrm{div}\mathbf{v}_h)\,\mathbf{I} - \mathrm{cof}[\nabla\mathbf{v}_h] \ . \tag{80}$$

Using (80) and integrating by parts we have

$$\begin{aligned} \textstyle\int_\Omega r\,(\nabla\mathbf{v}_h)^T : \nabla\mathbf{v}_h &= \int_\Omega r(\mathrm{div}\mathbf{v}_h)^2 - \int_\Omega r\mathrm{cof}[\nabla\mathbf{v}_h] : \nabla\mathbf{v}_h \\ &= \int_\Omega r(\mathrm{div}\mathbf{v}_h)^2 + \int_\Omega \mathrm{div}(r\mathrm{cof}[\nabla\mathbf{v}_h]) \cdot \mathbf{v}_h \ , \end{aligned} \tag{81}$$

where all the boundary integrals vanish because the function $r\mathbf{v}_h$ vanishes on the whole boundary $\partial\Omega$.

Due to Piola's identity and using (80) we have

$$\begin{aligned} \textstyle\int_\Omega \mathrm{div}(r\mathrm{cof}[\nabla\mathbf{v}_h])\cdot\mathbf{v}_h &= \int_\Omega \mathrm{cof}[\nabla\mathbf{v}_h]\nabla r\cdot\mathbf{v}_h \\ &= -\int_\Omega (\nabla\mathbf{v}_h)^T\nabla r\cdot\mathbf{v}_h + \int_\Omega (\mathrm{div}\mathbf{v}_h)\nabla r\cdot\mathbf{v}_h \end{aligned} \quad (82)$$

Recalling that on Γ we have $\nabla r = -\mathbf{n}$, simple algebra and an integration by parts give

$$\begin{aligned} -\textstyle\int_\Omega (\nabla\mathbf{v}_h)^T\nabla r\cdot\mathbf{v}_h &= -\int_\Omega \nabla r\cdot[\nabla\mathbf{v}_h]\mathbf{v}_h \\ &= \int_\Omega (\mathrm{div}\mathbf{v}_h)\nabla r\cdot\mathbf{v}_h - \int_\Gamma (\mathbf{v}_h\cdot\mathbf{n})(\mathbf{v}_h\cdot\nabla r) \\ &= \int_\Omega (\mathrm{div}\mathbf{v}_h)\nabla r\cdot\mathbf{v}_h + \int_\Gamma \mid \mathbf{v}_h\cdot\mathbf{n}\mid^2 \ . \end{aligned} \quad (83)$$

Using (81), (82) and (83) it follows

$$\int_\Omega r\,(\nabla\mathbf{v}_h)^T : \nabla\mathbf{v}_h = \int_\Omega r(\mathrm{div}\mathbf{v}_h)^2 + 2\int_\Omega (\mathrm{div}\mathbf{v}_h)(\nabla r\cdot\mathbf{v}_h) + \int_\Gamma \mid \mathbf{v}_h\cdot\mathbf{n}\mid^2 \ . \quad (84)$$

We now split each component of $\mathbf{v}_h$ in its linear and bubble parts, i.e.

$$\mathbf{v}_h = \mathbf{v}_h^L + \mathbf{v}_h^B = (v_1^L + v_1^B, v_2^L + v_2^B) \ . \quad (85)$$

Accordingly, noting also that

$$\nabla r\cdot\mathbf{v}_h = -(v_2^L + v_2^B) \ , \quad (86)$$

equation (84) can be written as

$$\begin{aligned} \textstyle\int_\Omega r\,(\nabla\mathbf{v}_h)^T &: \nabla\mathbf{v}_h = \int_\Omega r\big(\mathrm{div}(\mathbf{v}_h^L + \mathbf{v}_h^B)\big)^2 \\ &-2\int_\Omega \big(\mathrm{div}(\mathbf{v}_h^L + \mathbf{v}_h^B)\big)(v_2^L + v_2^B) + \int_\Gamma \mid (\mathbf{v}_h^L + \mathbf{v}_h^B)\cdot\mathbf{n}\mid^2 \\ &= A_1 + A_2 + A_3 \ . \end{aligned} \quad (87)$$

We now estimate the three terms above. We obviously have

$$A_1 := \int_\Omega r\big(\mathrm{div}(\mathbf{v}_h^L + \mathbf{v}_h^B)\big)^2 \geq 2 \int_\Omega r(\mathrm{div}\mathbf{v}_h^L)(\mathrm{div}\mathbf{v}_h^B) \ . \tag{88}$$

Integrating by parts and using that $\nabla r \cdot \mathbf{v}_h^B = -\mathbf{v}_2^B$, from (88) we obtain

$$A_1 \geq -2 \int_\Omega (\mathrm{div}\mathbf{v}_h^L) \nabla r \cdot \mathbf{v}_h^B = 2 \int_\Omega (\mathrm{div}\mathbf{v}_h^L) v_2^B \ . \tag{89}$$

Recalling that $\mathbf{v}_h = \mathbf{v}_h^L + \mathbf{v}_h^B \in K_h$ and observing that $\mathbf{v}_2^L$ is a continuous piecewise linear function, it follows that

$$A_2 := -2 \int_\Omega \big(\mathrm{div}(\mathbf{v}_h^L + \mathbf{v}_h^B)\big)(v_2^L + v_2^B) = -2 \int_\Omega \big(\mathrm{div}(\mathbf{v}_h^L + \mathbf{v}_h^B)\big) v_2^B \ . \tag{90}$$

Since it holds

$$\int_\Omega (\mathrm{div}\mathbf{v}_h^B) v_2^B = 0 \ , \tag{91}$$

from (90) we get

$$A_2 = -2 \int_\Omega (\mathrm{div}\mathbf{v}_h^L) v_2^B \ . \tag{92}$$

Furthermore, for A_3 we obviously have

$$A_3 := \int_\Gamma \mid (\mathbf{v}_h^L + \mathbf{v}_h^B) \cdot \mathbf{n} \mid^2 \geq 0 \ .$$

Collecting (89),(92) and (93), we finally get (cf. (87))

$$\int_\Omega r\,(\nabla \mathbf{v}_h)^T : \nabla \mathbf{v}_h \geq 2 \int_\Omega (\mathrm{div}\mathbf{v}_h^L) v_2^B - 2 \int_\Omega (\mathrm{div}\mathbf{v}_h^L) v_2^B = 0 \ . \tag{93}$$

The Proposition is proved. □

A.4 Proof for Proposition 4.3

Let us define the set $F(\gamma) \subset \Omega$ by

$$F(\gamma) = \big\{(X,Y) \in \Omega \ : \ 2\mu - \gamma\, r(X,Y) \leq 0\big\} \ . \tag{94}$$

Recalling that $r(X,Y) = 1 - Y$, the set $F(\gamma)$ has positive area for $\gamma > \mu$. It follows that, for sufficiently small h, there exists $\tilde{K} \in \mathcal{T}_h$ completely contained in $F(\gamma)$. Now take $(\mathbf{v}_h, \mathbf{E}_h) \in U_h \times S_h$ by choosing $\mathbf{v}_h = 0$ and $\mathbf{E}_h$ vanishing outside $\tilde{K}$. Moreover, in $\tilde{K}$ choose (cf. (44))

$$\mathbf{E}_{h|\tilde{K}} = \begin{bmatrix} \alpha\xi\eta & ; & 0 \\ 0 & ; & -\alpha\xi\eta \end{bmatrix} \qquad \text{with } \alpha \neq 0 \ . \tag{95}$$

Notice that $\mathrm{tr}\mathbf{E}_h = 0$; hence $(0,0) \neq (0,\mathbf{E}_h) \in K_h$ (cf. (46)). On the other hand, a direct computation shows that

$$a_\gamma(\mathbf{E}_h,\mathbf{E}_h) = \int_{\tilde{K}} (2\mu - \gamma r)|\mathbf{E}_h|^2 \ . \tag{96}$$

Since $\tilde{K} \subseteq F(\gamma)$, we have $(2\mu - \gamma r) \leq 0$, so that (96) implies

$$a_\gamma(\mathbf{E}_h,\mathbf{E}_h) \leq 0 \ . \tag{97}$$

As a consequence the form $a_\gamma(\cdot,\cdot)$ cannot be coercive on K_h for $\gamma > \mu$, uniformly in h. □

GPSR Compliance

The European Union's (EU) General Product Safety Regulation (GPSR) is a set of rules that requires consumer products to be safe and our obligations to ensure this.

If you have any concerns about our products, you can contact us on ProductSafety@springernature.com

In case Publisher is established outside the EU, the EU authorized representative is:

Springer Nature Customer Service Center GmbH
Europaplatz 3
69115 Heidelberg, Germany

Batch number: 10577876

Printed by Printforce, the Netherlands